MARX and ENGELS on ECOLOGY

Contributions in Philosophy

MARX and ENGELS on ECOLOGY

EDITED AND COMPILED BY

Howard L. Parsons

CONTRIBUTIONS IN PHILOSOPHY, NUMBER 8

GREENWOOD PRESS
WESTPORT, CONNECTICUT · LONDON, ENGLAND

83172

Library of Congress Cataloging in Publication Data
Main entry under title:

Marx and Engels on ecology.

(Contributions in philosophy; no. 8)
Bibliography: p.
Includes index.
1. Communism and ecology. I. Marx, Karl, 1818-1883.
Selections. English. 1977. II. Engels, Friedrich, 1820-1895.
Selections. English. 1977. III. Parsons, Howard L.
HX550.E25M37 335.43'8'30131 77-71866
ISBN 0-8371-9538-1

Library of Congress Catalog Card Number: 77-71866
ISBN: 0-8371-9538-1
ISSN: 0084-926X
First published in 1977

Greenwood Press, Inc.
51 Riverside Avenue, Westport, Connecticut 06880

Printed in the United States of America

Acknowledgments

I would like to acknowledge the permission of the following to reprint copyrighted and noncopyrighted material:

George Allen and Unwin Ltd. for selections from Frederick Engels, *The Condition of the Working-Class in England in 1844*, copyright 1968.

International Publishers for selections from Frederick Engels, *Herr Eugen Dühring's Revolution in Science (Anti-Dühring)*, ed. C. P. Dutt, trans. Emile Burns, copyright 1939 and 1966; Frederick Engels, *Dialectics of Nature*, copyright 1954; Karl Marx, *Capital: A Critique of Political Economy*, vol. 1, trans. Samuel Moore and Edward Aveling, and ed. Frederick Engels, copyright 1967; Karl Marx, *Capital: A Critique of Political Economy*, vol. 2, trans. Ernest Untermann and ed. Frederick Engels, copyright 1967; Karl Marx, *Capital: A Critique of Political Economy*, vol. 3, trans. Ernest Untermann and ed. Frederick Engels, copyright 1967; Karl Marx, *A Contribution to the Critique of Political Economy*, trans. N. L. Stone, copyright 1970; Karl Marx, *The Economic and Philosophic Manuscripts of 1844*, trans. Martin Milligan and ed. Dirk J. Struik, copyright 1964; Karl Marx and Friedrich Engels, *The German Ideology*, Parts 1 and 3, trans. and ed. R. Pascal, copyright 1947; Karl Marx and Frederick Engels, *Selected Correspondence 1846-1895*, trans. Donna Torr, copyright 1942.

Penguin Books Ltd. and Random House for selections from Karl Marx, *Grundrisse. Foundations of the Critique of Political Economy* (Rough Draft), trans. Martin Nicolaus, copyright Martin Nicolaus, 1973.

Progress Publishers for selections from Friedrich Engels, *The Housing Question* (1975); K. Marx and F. Engels, *The Holy Family* (1956); K. Marx and F. Engels, *On Religion* (1958); and Karl Marx and Frederick Engels, *Selected Correspondence* (n.d.).

Sanga Music Inc. for a selection from "My Rainbow Race" by Pete Seeger, ©. Copyright 1970 by Sanga Music Inc.

I am indebted to Mrs. Else Havanich for her able assistance in typing the manuscript. I owe thanks also to James T. Sabin, vice-president of Greenwood Press, Cynthia M. Harris, production editor, Betty C. Pessagno, copy editor, Mary Fremont, designer, and the reader of the original manuscript for their various contributions in bringing this book into being and in helping to improve its scope, organization, and expression. I must of course absolve them from any responsibility for the defects that remain and remain mine alone.

=====

The law doth punish man or woman
That steals the goose from off the common,
But lets the greater felon loose,
That steals the common from the goose.
  —Eighteenth-century protest against the
    Enclosure Acts in England, author unknown

One blue sky above us,
One ocean lapping all our shores,
One earth so green and round,
Who could ask for more?

And because I love you
I'll give it one more try
To show my rainbow race
It's too soon to die.
  —MY RAINBOW RACE by Pete Seeger

=====

Contents

Preface

When a colleague of mine, a director of studies on Shakespeare, learned that I was assembling the materials for this volume, he laughed and commented that I was now doing to the works of Marx and Engels what many scholars have done to the plays of Shakespeare.

He meant, I believe, that I was burrowing mole-like into the sources and would eventually emerge with a hodgepodge of bits and pieces on a particular topic—signifying nothing of consequence. The comment has a point, for the masters in various fields have written about and touched on a wide variety of things, trivial and significant; and to comb through the works of one (as is now being done by computers on the works of Shakespeare), compiling everything on which he declared himself on a particular topic, may itself be a trivial form of scholarship and an example of special pleading. For example, through this method pedants have tried to prove that Shakespeare was Christian and that Marx was an idealist or a misanthrope.

However, what Marx and Engels said on man, nature, and their relations to one another—in a word, on what today is addressed by the name of "ecology"—is a topic of quite another order. For them man is inconceivable apart from his evolution in nature and his collective labors upon nature by means of his tools. Man's dialectical relations with nature, in which man transforms it and is thereby transformed, is the very essence of his own nature. For man, nature is definable as the materials and forces of the environment that create man and are in turn created by man; and man is definable as a natural creator interacting with his environment. Thus, Marx and Engels had an understanding of an approach to ecology before the German zoologist, Ernst Haeckel, coined the term *Oekologie* in 1869, and long before the current "ecological crisis" and "energy crisis."

Today, ecology is one of the prime concerns of economic and political leaders and of people generally. The issues of production and distribution of natural resources, the shortage and uses of energy, pollution, population, food, poverty, and famine are featured in the newspapers almost every day. The scientific literature on these subjects, as the bibliography in this volume attests, is vast. The problems of our economy—inflation, prices, the cost of living, unemployment, underemployment, job insecurity, retirement, corporation profits—are pressing upon people with immediate urgency and increasing weight. Economy is a matter of ecology: it has to do with the production and distribution of goods and services in the context of human society and nature. The bread-and-butter issues are tied into a planetary system in which the peoples of the planet live and make their living by collective labors in and on and with nature. The growing ecological awareness is a recognition of this link between the price of bread and the total economic and ecological system. Eventually this awareness, driven by the logic of events and of reflection upon them, must lead to an acquaintance with Marxism's account of man's ecological situation, his ecological problem, and his ecological solution. It must lead people to ask why, under modern technology, so many produce so much, while under capitalism so many have so little of what is produced. It must lead them to ask why, under the ecological policies of monopoly capitalism, the natural environment is being destroyed along with the social environment.

Nation-states governed by the philosophy of Marx and Engels now embrace more than one billion persons on the planet. Because of the tremendous influence of this philosophy on these persons as well as on the millions living outside socialist societies who take that philosophy as "a living guide to action" (Lenin), it is important to make clear, both for Marxists and non-Marxists, where Marx and Engels stood on the principles and problems of ecology. There are both positive and negative reasons for defining their positions. The positive reason is that Marx and Engels, as the selections in this volume show, had a definite (though not fully detailed) ecological position. Political economy for them was inseparable from political ecology; as both working people and nature are exploited by class rule, so they will both be freed by liberation from class rule. The negative reason is that among many sections of the population, especially in the United States, much ignorance, misconception, and prejudice with respect to the teachings of Marxism-Leninism still prevail.

This volume unites the basic questions about ecology with the responses

given to those questions, so far as they have been given, by Marx and
Engels. Its purpose is to assemble the ideas of Marx and Engels on ecology
and, with the help of the philosophical framework and method they have
provided, to explore some of the important ecological issues, past and present.

The work consists of three parts: (1) an introduction; (2) selections
from the writings of Marx and Engels on ecology; and (3) a bibliography.
The introduction includes a brief survey of the historical background of
the thought of Marx and Engels, an exposition of their basic position on
ecology and of their critique of capitalist ecological policies, a statement
of the common criticisms of the Marxist ecological position and answers
to such criticisms, and, finally, a treatment of the problems of transition
from capitalist ecology to socialist ecology.

The selections, gathered from the great variety of Marx's and Engels'
works, include many but not all of their writings on ecology. They have
been culled and organized in order to exhibit with clarity and compre-
hensiveness their views on nature, man's interdependence with nature,
technology, the mutual transformation of humanity and nature through
labor, the precapitalist relations of man to nature, the capitalist ruination
of nature and of people in workplaces and dwellings, and the transformation of man's relations to nature under communism.

In the introduction, I have attempted to clarify the major ideas contained in these selections as well as the more general ideas of the philosophy
they presuppose. As particular notions of Marx and Engels are developed,
I have referred to the relevant passages from their work which are reproduced
in the selections. The notes to the introduction also refer to relevant parts
of their writings both here and elsewhere. The introduction is not merely
an exposition. Rather, I have endeavored to select, emphasize, and make
vivid the ecological notions of Marx and Engels essential to their general
outlook and of particular importance in the context of contemporary
ecological issues, theoretical and practical. I have sought to display the
ecological concerns of Marx and Engels—their concerns for humanity and
nature, their limitations, and their contributions to a humanistic solution
of ecological problems. In doing so, I have juxtaposed their ideas with the
ideas of many contemporary writers on ecology, bringing out the similar-
ities and differences.

The bibliography includes items not only in the fields of theoretical
and applied ecology as such but also in related areas, such as conservation,
economics, the environment, science, technology, and political, social, and

ethical thought. Ecological literature today covers a very large field, and the bibliography has been selected with regard to the issues raised in this work.

Marx and Engels never claimed to be infallible or exhaustive on these questions or any others. But they did employ a very powerful method of analysis and criticism which, together with the revolutionary practice and socialist commitment of millions, has transformed the modern world. Nor could Marx and Engels have foreseen the particular problems of the late twentieth century—nuclear armaments and the possibility of nuclear war, the evils of fascism and racism, the danger of rising famine and disease, and the threat of ecological disaster. Marx wrote in his Preface to the first volume of *Capital*: "I pre-suppose, of course, a reader who is willing to learn something new and therefore to think for himself." To read the work of Marx and Engels openly and carefully is always to learn something new. But the ultimate question for us the living is whether we can solve the problems that we must solve if we and our descendants on this planet are to live and to live in fulfilled ways. Hence the questions that we must put to the ideas of Marx and Engels, as to any other ideas, are: Are they relevant to the solution of such problems? Can we develop them, improve upon them, and usefully apply them to our problems in ways that help us to understand and to solve those problems? This work is offered in the conviction that the ideas of Marx and Engels on ecology are and can be made relevant and useful in solving ecological problems and the other major social problems connected with them.

part 1
Introduction

I. Background to the Thought of Marx and Engels

The position of Marx and Engels on ecology embraces their position on technology, for they understood man as a natural being in dialectical interpenetration with the rest of nature by means of his perceptions, his reflections, his manipulatory practice with tools, machines, and techniques, his consuming, and his enjoyments. As nature and the practice of man reciprocally call out and influence each other, so the concepts pertinent to nature and human techniques—i.e., the sciences of ecology and technology—must be reciprocally advanced.

The social relations of men to one another determine their concept of external nature, and in turn their relations of living and livelihood toward nature determine their social relations and their concepts of both society and nature. Paleolithic men, organized into small, intimate, cooperative groups in tension with other groups, obtaining their food through the use of primitive tools such as crudely shaped sticks and stones, appear to have conceived of nature in a simple, dialectical manner: man was seen to be identical with nature in a dumb, undifferentiated way, yet nature stood over against him as an alien, mysterious force, sometimes benign, sometimes destructive. Neolithic, slave-owning, feudal, and capitalist societies have all had their characteristic concepts of nature, each reflecting the prevailing mode of material production of the society: maternal, despotic, tightly hierarchical, and atomic. In contemporary times, all of these concepts may still be found as remnants from the past, adopted in response to tradition or in order to suit personal purposes or social circumstances. In these concepts, man's relation to nature is variously viewed as a mother-child relation, in which a supreme, personal, sacred power dispensing goods and evils can be appeased by man, e.g., in the religions of many parts of India and Africa, and in Mahayana Buddhism; as a despot-slave relation, wherein the supernatural despot arbitrarily disposes of his property (the things and persons of the lower order of "nature") and strictly rewards and punishes his subjects, e.g., in Islam and orthodox Christianity; as a compact, hierarchical system, ranging from inanimate beings through personal beings to the supreme spiritual being which orders all of nature according to his purpose, e.g., Christian feudalism; and as an atomic-mechanical system, devoid of values, purpose, and spirit, determined by the laws of exchange, e.g., capitalism.

The truly modern concept of nature is the *ecological* or *dialectical*. It was no accident, as we shall see, that ecology and dialectics both arose in the nineteenth century in response to the revolutionary achievements in societies and the sciences, especially the biological and historical sciences, which revealed the interactive, evolutionary, and transformative character of nature and society. Thus today, in the late twentieth century, the fastest growing concept of nature among all those extant is the dialectical concept. This concept is implicit in the very method and findings of the sciences, and while it is not always explicit and conscious, it is widely shared in the thought and practice of scientists the world over. Moreover, one-third of the world's people are citizens of socialist social systems in which the dialectical method for understanding nature and man is taken to be definitive. Hence, the concepts of nature that dominated men's minds in the past and still vie for their allegiance are being rapidly displaced by this new concept.

Dialectics is the study and description of the fundamental categories of nature. Dialectics rejects initially—for reasons provided by both human practice and theoretical inquiry—a supernatural realm and a static world of individual entities, i.e., the metaphysical, mystifying ideologies of feudalism and capitalism. It takes "nature"—the world of material things and their relations assuming the forms of mass and energy, substance and field, and so forth—to comprise all past, present, and future existence, all actuality and potentiality, all being and becoming, all unity and diversity, all quantity and quality, all levels of material existence ranging from the most elementary to minded and "spiritual" entities. Dialectics applies to social phenomena, such as revolutions, no less than to quanta. It asserts the ultimacy of spatio-temporal events interlocked in the context of other events—processes (energies) that arise out of a background of other processes, interact with other processes, unite and conflict with them, change and are changed by them, become, develop, decline, and pass away.

This process of becoming, of mutuality, unity, and opposition of inter-action with other processes, is marked by several features. It is vibratory and pulsating. The continuity of its pulsations is a continuity of repetitive steps with graduated changes, at more or less complex stages. Such changes occur in small quantitative steps that eventually may issue in sharp qualita-tive leaps of change—e.g., the birth or death of a human being, or the death of one social order and the revolutionary creation of a new one. Every process is in some degree a negation of those processes with which it inter-acts and on which it depends for its own identity-career; this identity-career in turn is in process of being negated. Its stable form arises from this interdependence, but the contradiction between the process and its opposite processes is more basic than the unity. The unity of individuals and interindividual relations is relative and transient; their opposition and transformation are absolute and abiding.

The laws of dialectics—and we have indicated only a few of the basic ones now known—obtain for all kinds and levels of material existence, as well as for the interactive process and the creative labor of human beings with the objective world. In this process, by which we live and make a living, we acquire, apply, and repeatedly test our knowledge. (See Selections IA; II; III.)

Dialectical activity and regularity are the basis for all relations of organisms with other organisms and their physical environments. Such relations presuppose the existence of a certain physical and biological order, created, sustained, and transformed by dialectical processes. A basic model for the understanding of such order is the concept of the *field*. A field is "a system of order such that the position taken up by unstable entities in one portion of the system bears a definite relation to the position taken up by unstable entities in other portions."[1] The entities may be subatomic particles or stars, and in both their internal structures and their external relations they provide a space-time frame for the forces constituting and affecting them. A field is a region of energetic activity differentiated into particular "individuals" integrated into a general order that may be qualified by local orders. Every individual is in turn a field of subindividuals, and at present writing no "ultimate" individual or partless particle is known to science. Individuals are in some sense discontinuous (quanta are striking instances of discontinuity) but are related to other individuals by the continuity of physical inheritance from the past and of lines of forces in their fields binding them together in space-time regions.

While the general order and individuals of fields require and affect each other, their relations to one another vary. If the order is changed, the individual may be greatly affected (organism, atom) or little affected (chemically bonded atoms); if the individual is changed, the order may be greatly affected (two persons in love) or little affected (a person in a large impersonal institution). The order may endure when the individual is destroyed or displaced (primitive organisms that regenerate parts), and the individual may endure when the order is destroyed or displaced (atoms in molecules, persons in families, institutions in societies).

The difference between a physical field and a biological field is that in the latter the structures of the individual entities—genes, cells, organisms, species, interspecies relations—function to facilitate energy exchanges[2] among living entities and are therefore relatively transient. Such living individual entities are mortal. But in their temporary roles they contribute to the unfolding drama of a system; moreover, provision is made for their replacement. Whereas in physical systems the individual entities are locked into a relatively rigid relation with one another and the system, in living systems a balance is struck between the equilibrium of the system and the flexibility of the parts. There is a gain in diversity, adaptability, and development of the system, but the price is instability and mortality in the parts, a price not exacted in the safety of physical systems. For example, the quartz silicon crystals of a granite rock have relatively permanent internal structures and external relations, and there is little exchange of energy between the crystals and the environment. The crystals are durable but undiversified and undeveloping. They do not "live for" the system in the sense of actively contributing to the endurance and development of the system. By contrast, an organism contributes to those relations that maintain and extend its species and its species-environment relations. Because it is an open system of energy, easily giving and taking, it soon dies, to be succeeded by a replacement. The individual organism passes, but the system defining the species goes on.

The production, storage, and expenditure of energy in living systems— in contrast to the high entropy of physical systems—increase the probability of novelty and diversification. Physical systems are relatively simple, stable, and resistant to change. Novelty in entities means an increase of complexity, instability, and change. Living systems with their free energy and their disequilibrium are receptive to novel entities, though the range of novelties acceptable and assimilable is limited, and the time required for the integration of the novelties may be long. Diversification in a living system has adaptive value of course, but it also often seems to be the product of sheer exuberance and expressive fancy.

From this perspective, plant and animal organisms are in mutual, continuous, and transformative relations with their immediate environments, in a universe of natural systems of things and events in dynamic equilibrium. Strictly speaking, ecology is concerned with the interactions of two or more complex living systems with one another and with the physical environment.[3] Such subject matter—an ecosystem—is only a portion of the total universe of nature, so that one may study nature without bringing in ecology. On the other hand, ecology as a specific science of ecosystems displays the principles of the general science of nature, or dialectics. For dialectics as a science of systems generally is concerned with the interactions of two or more living or nonliving systems with one another and with their environment. Ecology is the application of dialectics to living systems, and dialectics is the generalization of the method of ecology from living systems to all systems.

The perspective of nature as a unitary and dynamic system of systems arose as the result of the convergence of some germinal ideas in the natural sciences at the end of the nineteenth century: the notion of ubiquitous energy (atomic theory, field theory, the conservation of energy, the equivalence of matter and energy); the notion of the continuity of energy and life (cell theory, organic chemistry); and the notion of the creation and evolution of forms of living energy (evolutionary theory). In this perspective every thing or event is an energetic process or nexus of processes defined by its interconnections with its near neighbors (including its immediate past) and by indirect interconnections with its remote neighbors. Nineteenth-century concepts of nature which expressed this convergence and this unitary, evolutionary view, such as the materialism of Ernst Haeckel (1834-1919), were materialistic. They emphasized the *physical* unity of nature, and they acknowledged the fact of the *biological* and *ecological* unity in the broad setting of physical nature. (Haeckel invented the term *ecology* a century ago.) But they tended to subordinate biological to physical categories. This subordination evoked the common criticism that materialism is "reductive" and antispiritual.

Through the Middle Ages to the late eighteenth century, the prevailing concept of nature was that of "the great chain of being"—a scaled, continuous universe of discrete, settled things in a *plenum formarum*.[4] In the nineteenth century, the concept took on the notion of strife, interpenetration, and transformation among things. The rapid expansion of large-scale industry and great advances in science and the technological power of men to know, predict, and control the world of nature indicated a nature whose orders were subject to man's intervention and change. The spinning jenny, power loom, cotton gin, sewing machine, steam hammer, steamboat, locomotive, camera, achromatic compound microscope, spectroscope, ophthalmoscope, telephone, telegraph, phonograph, electric light, anesthetic and antiseptic surgery, vaccination, controlled fermentation, internal combustion engine, Bessemer process, manufacture of dyes, nitroglycerine, vulcanization of rubber, discoveries in zoology, physiology, geology, and paleontology—all were proofs of this power. Accordingly, the feudal concept of nature was undermined and the way was prepared for a new, dialectical interpretation. Marx and Engels formulated only the outlines of this new concept and did not fully work them out; Marx wrote little on the subject, and though Engels wrote *Anti-Dühring*, he never completed *Dialectics of Nature*. (See Selection II.)

II. Marx and Engels on Ecology

The significant feature of the work of Marx and Engels in this connection was not their sense of the unity of the whole (German, archaic, feudal romanticism and nature philosophy had this sense) or their dialectics (Hegel had this). Rather, it was their appreciation of the dialectical power in human history and society and their grasp of the dialectical effects of social practice upon the world of external nature. Marx and Engels were the first major scientists of human society. Hence, as the scientists of nature of the nineteenth century pictured nature under the model of mechanical materialism, making society derivative from the forces of inorganic nature, Marx and Engels, true to their German background, concentrated upon society, with its actual and potential power over nature. Then as now in capitalist countries, the natural scientists tended to be apolitical or else conservative,[5] whereas the "natural history" of the German school, while it had its reactionary devotees, could be creative and optimistic in the face of the drab regularities of the Newtonian mechanistic universe. Indeed, it was the viewpoint of an expanding, conflict-ridden, innovating, problem-confronting, and problem-solving society which provided a prime stimulus and root metaphor for the concepts of nature in Hegel, Darwin, and Marx. Capitalism, as Marx correctly described it, was an omnivorous power, sweeping all before it. It was engaged in "constant revolutionizing of production," and like Siva it destroyed what it created. It was the very embodiment of Hegel's *Weltgeist,* which perpetually objectifies and then cancels itself.

Marx and Engels were aware that as man is the evolutionary product of natural processes, so man as a social being acts back on nature to change it and direct it according to his purposes. Though they held to a unitary, dialectical concept of nature, they developed a *social* or *human* science not only because they wished to combat the conservative effects of mechanical materialism, supernaturalism, and idealism, but also because as radicals they desired to affirm the creative role of man in shaping history and nature. Yet *social* man is always conceived as a *natural* being, and external nature is always conceived as the other side of man, both individual and social, which is constantly humanized in man's dialectical dealings with it. But the difference between Marx and Engels, and many "natural scientists," is that science for the natural scientists is conceived and practiced abstractly, i.e., separately from generic human needs and problems, whereas for Marx and Engels it is from start to finish a human enterprise, a creatively humanized nature in Nature. Marx wrote:

> *Industry* is the *actual,* historical relationship of nature, and therefore of natural science, to man. If, therefore, industry is conceived as the *exoteric* revelation of man's *essential powers,* we also gain an understanding of the *human* essence of nature or of the *natural* essence of man. . . . One basis for life and another basis for *science* is *a priori* a lie. (See Selection X B.)

In this sense man in his universal development is "able to understand his own history as a *process,* and to conceive of nature (involving also practical control over it) as his own real body." Man "annexes" nature to his own bodily organs. His labor is the interaction of man and nature wherein "man of his own accord starts, regulates, and controls the material reactions between himself and Nature." And "by thus acting on the external world and changing it, he at the same time changes his own nature." Marx stresses the role of man's own imagination, purpose, and will in this inter-action, and the power of man to put the stamp of his value-creating labor on nature's materials and processes. (See Selection VI C.) Yet, the sense of nature as a prior objective power conditioning man is there, the sense that while man "opposes" nature he is always *with* it. Man makes nature do his bidding, acquires command over it, and even causes the product of his past labor, machinery, to "work on a large scale gratuitously, like the forces of Nature." (See Selection VIII A.) But man can do all that only because his own thought, will, and practice are embedded in and organi-cally integrated with the processes of nature. Plants and animals, for ex-ample, originally provided by nature as means of subsistence and material for labor, become through man's laboring intervention instruments of production, transformed and being transformed by man's life for genera-tions. (See Selection VI C.)

For Marx, man participates organically, i.e., dialectically, in nature. In 1844, Marx described nature as man's *"inorganic body"*—that "with which he must remain in continuous intercourse if he is not to die." Plants and animals, stones and air and light are not only "the material, the object, and the instrument" of man's practical activity for obtaining food, clothes, shelter, heat, and other necessities. They are also objects of man's "spiritual" nature—we may say both the stimuli and the projective locales for man's science and art—and it is through them that man comes to the full realization of himself. "That man's physical and spiritual life is linked to nature means simply that nature is linked to itself, for man is a part of nature." Has there been a better, a more succinct, ecological statement of man's place in nature than this? (See Selection III A.)

Marx arrived at this dialectical, ecological standpoint in refutation of Hegelian idealism, wherein the world of nature is alienated from man's self-consciousness, and in opposition to atomic materialism, wherein nature is conceived to consist of particulate individuals, discontinuous and dead. As a natural being, Marx argues, man is both active and receptive. He requires objects for his impulses, as objects in turn fulfill his needs, such as hunger. Moreover, Marx generalizes this relation of mutuality so that it obtains, he says, for the sun and the plant: the sun is "confirming" the life of the plant, while the plant is an "expression" of the power of the sun. (See Selection III B.) Marx never repudiated this understanding of the mutuality in nature, though after his early Paris period he would have used designative rather than appraisive language for signifying the sun-plant relation.[6] In the *Grundrisse* and *Capital,* he used the scientific term popularized by Moleschott, *Stoffwechsel,* variously translated as "exchange," "circulation," and "metabolism."[7] Moreover, the mutuality is conceived dynamically, so that while here, in opposition to Hegel, the creation and *balance* of nature are described, the unbalancing and destruction also occur. Marx interprets history in this dialectical way while not working out the dialectics of nature as such.

For Marx, nature prior to man's actions provides man with natural wealth in the form of means of subsistence (fruitful soil, fish, etc.) and in the form of labor instruments (waterfalls, navigable rivers, wood, etc.). The physical conditions of nature also fetter him. If man's wants are few, the soil is fertile, and the climate is favorable, then labor-time will be minimal. But if the soil, its products, and the climate (temperate) are differentiated, man's wants and abilities are also differentiated. Necessity becomes the mother of inventiveness in production, as in the development of irrigation works. Thus the mode of advanced production, capitalism, "is based on the dominion of man over Nature," and the temperate zone "is the mother-country of capital."[8] (See Selection I E.)

Engels' work on *Dialectics of Nature* came relatively late; he first outlined it in 1873. It was an effort to generalize the findings of the sciences in a materialist, dialectical way. During his youthful Paris experience, however, Marx (together with Engels) was drawn into practical, political activity and thence into the analytical work of political economy. Hence, the emphasis on an ecological naturalism or materialism, which is strong in *The Economic and Philosophic Manuscripts of 1844,* was developed in the form of a study of the problems of man in society and their economic, political solutions. Marx increasingly focused on the active, practical role of men in shaping nature in accord with their needs and purposes. The ecological materialism was never abandoned. Marx retained the idea of man in interaction with the rest of nature but stressed man's own self-initiated political economy in shaping nature and society.

The "unity of man with nature," the young Marx observed, has always been expressed through man's industry toward nature, alongside the struggle of man with nature.[9] The unity is qualified and developed through the struggle. For the prime fact of man's life is his hunger and other imperative needs, driving him to act toward nature and to exchange his labor for the products of nature.[10] Nature exists prior to man's trans-

formative labor upon it. (See Selections I A, B, D; VI B, C, G; VIII A, D.)
It "spontaneously" provides fish, timber, and ores, which man can easily
detach from their connection with their environment. But in the form of
stones it also provides man with tools "for throwing, grinding, pressing,
cutting, etc." Strictly speaking, man's first tools were not these stones but
his "own limbs" which served as instruments for obtaining his means of
subsistence. (See Selection VI C.) Such ready-made instruments are
natural technology, and Marx points out that Darwin has "interested us
in the history of Nature's Technology, i.e., in the formation of the organs
of plants and animals" serving as instruments for producing and sustaining life. (See Selection V A.)[11] Since Neolithic times man has employed
animal power and abilities (e.g., milk production) for his own purposes.
His most primitive Paleolithic stone tools were probably originally copies
from the cutting and scraping teeth and claws of animals, the piercing
bills of birds, and the like. Man, of course, through the use of his brain
and hand, extends the power of his bodily organs by the construction of
tools and machines. These are "natural material transformed into organs
of the human will to dominate nature or to realise itself therein."[12]

All human technology thus arises out of natural materials, on a natural
base, and in a natural context as an effort of man to survive and to realize
his powers. If technology along with the world it creates is understood
to be uniquely different from nature, a jinni or Frankenstein monster
that will destroy its maker, then it is misunderstood.[13] If technology is
thought to be incurably "alienated" from nature, so that the two are ir-
reconcilable, like the "two cultures," then it is misunderstood. For if
technology is the product of the brain, hand, and will of man, and has
gone wrong, then it is man himself who has the power to observe its
effects in nature, to study it, and to rectify it. Modern ecology shows
that technology is no mystical monster but has specific, tangible effects
on the materials and processes of nature which can be known and controlled. We know, for example, that the smog pollution in most American
cities is the result of chemicals triggered by nitrogen oxide in the air; that
this nitrogen oxide is the consequence of higher temperatures caused by
bigger compression engines designed for bigger cars in the last thirty years;
and that the solution is to bring public pressure on car manufacturers to
reduce the size of cars and hence the size of profits.[14] We know that the
big manufacturers discharge untold tons of pollutants into the air and
megatons of sludge into the rivers, lakes, and oceans. We know that the

food processers pack our food with poisons. We know that the powerful airlines pour out pollution and intolerable noise into public air space. We know that the military-industrial complex sowed its seeds of herbicidal destruction and soil ruination in Vietnam. We know that these are big monopolies, laws unto themselves, responsible to no one. And we know that they do what they do for profit and superprofit. Is "technology" the culprit? The technician? The people? ("We have met the enemy, and it is us" is a self-accusing and self-absolving moralism, and irresponsibly lets the Big Enemies off.) "System dynamics"? (Jay W. Forrester). "Propensity for growth"? (D. H. Meadows). "The tyranny of progress"? (Robert Gower). "The ecological asymptote"? (Kenneth Watt, Harvey Wheeler). These abstractions sound like the incomprehensible mumbo-jumbo of ancient religions. Gus Hall hit the real nail on the real head: "The ones responsible are the monopolies, the corporations, who have made enormous profits while they pollute and destroy our environment. . . . The major victims are the working people, who suffer from the effects of pollution every minute of their lives."[15]

If we are to describe the problem accurately, giving due weight to its dialectical complexity, we cannot simply say that the monopolies are the only ones responsible. Individual monopolists, i.e., owners of large and dominant industrial corporations and agribusinesses, comprise only a very small proportion of the population. They stand at the top of an economic and social pyramid that embraces the aggregation of capital and machinery, of technological knowledge, skills, and processes, and of executives, managers, technicians, and workers of all kinds. Around this pyramid of productive forces arises a network of productive relations into which men enter as they produce and exchange goods and services. Thus the pollution of the environment (through, for example, the pollutants of combustion or the discharge of noxious substances onto land or into waters) is a built-in consequence of the whole system as it is organized, and all who are engaged in production and distribution are, therefore, also engaged in pollution. In general terms, the whole productive system is the polluter.

Does this mean that all individual persons and groups in that system are equally responsible? By no means. Given the capacity for equal reasonableness among men, responsibility is generally proportionate to power: the greater the power, the greater the responsibility. In a capitalist system, the monopoly of power, or ruling control, rests with the big capitalists, i.e., the monopolists, allied with the government and the military. (Eisenhower used the term "the military-industrial complex.") The central motive force of this system is profit-maximization, and one consequence of this force is environmental pollution. It is a historic consequence in the sense that pollution seems to follow, causally and invariably, from a system of profit-maximization and a logical consequence in the sense that when the capitalist is confronted with a long-run choice, rather than a particular choice, between profit and environmental pollution, he must in all logic choose profit. To choose pollution and deny profit would be to deny his own definition and existence. Thus, so far as the capitalist class exercises ruling control in the system, a control arising out of its direction of the central motive force, it is the most responsible class for what occurs in the system.

But the workers also constitute a motive force, secondary to the motive force of the capitalists, a force of a different kind, a productive force. So long as the power of the working class is subordinate to the power of the monopoly class, the working class, both as producer and as consumer, has a subordinate responsibility for the consequences of the system. But the working class is not merely a static and secondary class, fixed forever to subordination under a ruling class. It is potentially a dominant and universal class, and it is moving in the direction of throwing off its subordination and becoming just that kind of universal class. It is, therefore, a class responsible for doing what it does within the existing class system, responsible for changing its position within that system by becoming progressively more class-conscious and hence more class-responsible, and a class responsible for assuming a position of ultimate power and responsibility over the whole transformed system.

The people or workers, in short, being integral parts of a system that has exploitation of man and nature built into its center, are partly responsible for the consequences of that system but not primarily responsible, for they do not determine the central motive force and primary decisions of the system. They are secondarily responsible insofar as they exercise some power within the system and have the ability to obtain facts, use reason, and take action that can change the system into a better one. They possess the potential power for a revolutionary transformation of that system. In the long run it is the people who must assume the ultimate responsibility if exploitation of man and nature is to be progressively reduced and eventually eliminated.

Solving the problem of exploitation in the existing socioeconomic system is not an either-or choice between, on the one side, small reforms, such as conserving electricity in the home, or large reforms, such as immediate and heavy penalties for industrial owners who pollute the environment, and, on the other side, a total transformation of the system into a truly democratic, socialized society. Small reforms can lead to larger and larger reforms, as changes in the social consciousness of individual persons can lead to changes in the consciousness of people in groups, institutions, and the whole society. Such changes in consciousness carry with them changes in action, and scattered group and institutional actions can eventually culminate in new quantitative and qualitative changes that will issue in a new society, eradicating class exploitation and the differentiation of classes themselves.

In his adjustive activities toward his environment, whether the rural world of soil, plants, and animals or the urban world of finished products, the individual person passes through the phases of receptivity and dependence, active dominance, and detachment and consummatory enjoyment. Some human societies have passed through similar phases: primitive identification with and dependence upon nature, technological conquest of nature, and comfortable enjoyment of it and its products.

Marx and Engels viewed Western capitalism, with its progressing technology and "constantly expanding market," as supplanting the passive barter economy of feudalism. Capitalism aimed at subjugating nature, whereas feudalism obeyed and passively acquiesced in it. For capitalism

> for the first time, nature becomes purely an object for human-kind, purely a matter of utility; ceases to be recognized as a power for itself; and the theoretical discovery of its autonomous laws appears merely as a ruse so as to subjugate it under human needs, whether as an object of consumption or as a means of production. In accord with this tendency, capital drives beyond national barriers and prejudices as much as beyond nature worship, as well as all traditional, confined, complacent, encrusted satisfactions of present needs, and reproductions of old ways of life.[16]

Just as modern capitalism has destroyed "feudal, patriarchal, idyllic relations" with "egotistical calculation," "resolved personal worth into exchange value," and depersonalized and desacralized personal and occupational honor with "naked, shameless, direct, brutal exploitation"[17]—so it has plundered and smashed that dumb, slumbering, untapped feudal world of nature and with venal vigor has plowed anew its soil, fished its waters, felled its forests, probed the depths of its earth, built new ships, dredged new harbors, and where only quiet towns stood has reared cities of smoking factories and wretched tenements. Under nascent capitalism, nature was no longer a design to draw solace and guidance from; it was a mere inert, defenseless thing to be seized, shaped, consumed, and discarded. It was no longer a limited translucent order through which an infinite God was to be worshipped; it was the raw material, the unbounded storehouse for feeding the voracious appetite of man, i.e., capitalistic man.

Marx and Engels discerned in this development of the new economy two social movements contradicting each other. One was the "subjection of nature's forces to man,"[18] the expansion of man's productive forces to meet his material needs. Such an "enlarging of the bounds of human empire," as Francis Bacon called it in 1629 in his *New Atlantis,* would open the way to true human freedom. The other movement was the inertia and failure of the existing system of social relations with respect to controlling the new power and wealth in the interests of the vast majority. When Marx and Engels said that "the weapons with which the bourgeoisie felled feudalism to the ground are now turned against the bourgeoisie itself,"[19] they meant that the productive forces generated by capitalism— the productive apparatus of technology and the proletariat who oppose capital—were in process of revolting against the fetters that prevented their full development. They also meant that the ecological policies of capitalism in its exploitation of nature, in contradiction to the demands and laws of an objective and resistant nature, must be understood and overcome by that same class, the proletariat. The world is lawful. An economic system that fails to secure the laws of fulfillment among human beings is destined to generate revolt by human beings. In a parallel way, an economic system that breaches the laws of nature by which wealth is produced will bring on inevitable reactions: impairment of nature's "metabolism" of ecological cycles, depletion of nature's resources, impoverishment of human society, and a relapse of nature into the slumber of undevelopment. In *The Deserted Village,* Oliver Goldsmith portrayed in poignant and bitter tones the consequences of the eighteenth-century transition from an agrarian economy to the raising of sheep and the wool trade—the enclosure laws, the disappearance of the freeholder, and the

displacement of the peasant from the land and his forced migration to
the city. Vividly he bewailed the sickly barrenness of the land and the
famine-stricken peasants alongside the "sickly greatness" of the luxurious
ruling class:

> Ill fares the land, to hastening ills a prey,
> Where wealth accumulates, and men decay.

Early, Marx wrote of this commercial revolution in England: "It destroyed
natural growth in general as far as this is possible while labour exists, and
resolved all natural relationships into money relationships."[20] And later:
"The spoliation of the church's property, the fraudulent alienation, the
State domains, the robbery of the common lands, the usurpation of feudal
and clan property, and its transformation into modern private property
under circumstances of reckless terrorism, were just so many idyllic methods
of primitive accumulation."[21] Marx noted in his own day the same inconsistencies in this system that Goldsmith had recorded, though more precisely: "the cleanly weeded land, and the uncleanly human weeds, of
Lincolnshire, are pole and counterpole of capitalistic production." (See
Selection VIII J.)

The failure of capital to control itself is thus manifest in the domains of
both human relations and man's relation to nature. It not only divides
city from country, concentrates in cities men as well as matter necessary
to the soil, and imperils and impoverishes the industrial and rural workers.
It also robs the soil as well as the laborer, both of them "the original
sources of all wealth."[22] At the same time it "imperiously calls" for the
"restoration" of the ecological circulation of matter "as a system" necessary to the "lasting fertility of the soil" and in a form "appropriate to
the full development of the human race." (See Selections VIII D, A.)
Thus, as man and nature are conjointly exploited by capital, so it is man's
mission to recover the "naturally grown" unity of nature *in conjunction*
with his own unitary fulfillment.

Concentrating on the contradiction in society, Marx and Engels did
not discuss extensively the contradictory relations of man with nature.
Yet, they had the latter continuously in mind, as their criticisms of capitalism's exploitation of nature demonstrate.

"The mode of perceiving nature, under the rule of private property
and money," Marx wrote in 1844, "is a real contempt for, and a practical

degradation of, nature."[23] He cited the ecological protest of Thomas
Münzer against the private ownership of fishes, birds, and plants. Under
private property, man relates himself to nature in the mode of *possession,*
of *having* and *grasping.* (See Selection X B.) Such a mode is denaturalizing
and dehumanizing. Man needs objects beyond him to activate and unfold
his powers. Therefore, to deny (as in idealism) or to dominate greedily
(as in vulgar materialism) the objects of the external world is to try to
escape what we vitally depend on for our active, sensuous, thoughtful
development, namely, objects "indispensable" to the "confirmation" of
man's powers. (See Selection III B.)

The attitude and practice of man toward nature in capitalism have given
rise to a peculiar split in man's individual and social life. It has separated
nature from man, industry from humanization, work from play, science
from art. Nature is thought to be dead, neutral, discrete, movable bodies,
indifferent to consequences and values, submissive to man's whims and
demands, without any structures and demands of its own not directable
by man, and without any responsiveness. The primary science from the
seventeenth to the nineteenth centuries was mechanics, which dealt with
just such bodies. (The attitude implicit in such a science was not only a
nineteenth-century attitude; it is widely shared today by many business-
men, politicians, urban planners, real estate developers, and urban residents.
Yet, nature exists independently of man's thought; it was here prior to the
human species and each individual person, and it has its own objective
laws that govern us and that we violate at our own peril.

Marx was cognizant of this last fact. He wrote of the exhaustion of the
soil by industry and commerce (see Selection VIII G), just as he wrote of
the exhaustion of the individual laborer and of the wastage of nature's
possibilities as of human talents.[24] Man and nature mutually condition one
another. Discussing the factors that limit the productivity of labor, Marx
mentions social factors, such as "anarchy of competition" and "the
peculiarity of the bourgeois mode of production," as well as "natural
premises." The natural premises "frequently become less productive as
productivity grows—inasmuch as the latter depends on social conditions."
Thus "the exhaustion of forest lands, coal and iron mines, etc.," which
is the result of man's productivity, diminishes the development of that
very productivity.[25] The capitalist who simply seizes the living labor of
men and pays nothing in return for a portion of it likewise seizes the
physical materials and forces of nature; the latter "cost . . . nothing."

(See Selection VIII A.)[26] As the wanton seizure in the first instance is a violation of the laws of human needs,[27] so the seizure in the second instance results in a violation of natural laws because the same attitude that produces the seizure prompts the appropriation and misuse of natural resources.

Capitalist agriculture, Marx remarked, "works against a rational agriculture" by, for example, removing rural workers from the land.[28] For capitalism, quick profit from land is more important than the welfare of human beings or even the maintenance of the land's fertility. It simultaneously "robs" the soil of the return to it of the elements (in waste form) consumed by man, namely, food and clothing. It also isolates and demoralizes the rural worker, "crushing out" his "vitality, freedom, and independence." (See Selection VIII D.) Violation of human nature is *ipso facto* violation of nonhuman nature, since man is an organic part of a wider nature. Capitalism, which both dehumanizes man and denatures nature, thus needs a humanistic, naturalistic alternative: "the control of associated producers."[29]

The movement of landed monopoly, Marx observed, drives men into crowded industrial cities. "It thereby creates conditions which cause an irreparable break in the coherence of social interchange prescribed by the natural laws of life. As a result, the vitality of the soil is squandered, and this prodigality is carried by commerce far beyond the borders of a particular state (Liebig)." (See Selection VIII G.)[30] Here Marx had in mind the importance of manure to plant life, as stressed by Liebig, the agricultural chemist.[31] Marx wrote of how wastes—he called them "excretions of production and consumption"—are utilized. (See Selection VIII E.) But he remarked how the "exploitation and squandering of the vitality of the soil . . . takes the place of conscious rational cultivation of the soil as eternal communal property, an inalienable condition for the existence and reproduction of a chain of successive generations of the human race." Preservation of this "eternal" ground of the human species, of this "inalienable condition" of mankind's future, requires a conservation policy: "the rational cultivation, maintenance and improvement of the soil itself." (See Selection VIII G.) In these passages, one should note Marx's sensitivity to the objective existence of ecological laws, to man's participation in and violation of them, and to man's responsibility to them as a form of his responsibility to himself and to future generations. Here the true humanist turns out to be the true naturalist, and the true radical to be the true conservative.

Engels had some incisive things to say about the narrowness of natural scientists[32] and the "revenge" that nature takes on us for "our human conquests over nature" and the "unforeseen effects" of these conquests. He cited the destruction of forests in Mesopotamia, Greece, and Asia Minor, and the consequent and unrealized removal of water reservoirs; the exhaustion of fir forests by Alpine Italians and the consequent impairment of the dairy industry and mountain springs, and the creation of furious floods; and the spread of scrofula following the wide use of the potato in Europe. Moreover, Engels said, individual capitalists are concerned only with the immediate and profitable effects. For example, the Spanish planters in Cuba burned down mountain forests to get ash fertilizer for one generation of profitable coffee trees but produced floods that washed away the upper stratum of soil. Engels further observed how the introduction of goats into Greece prevented the regeneration of forests there, and how the goats and pigs brought to St. Helena radically changed the vegetative ecology of the island. (See Selection VIII F.)

Given their political interests and the stage of bourgeois science, Marx and Engels were bound to concentrate on the devastations of capitalism in the city, though corresponding devastations in the countryside did not escape them. For example, Marx said that large industry "lays waste and destroys" the labor-power of men, while large agriculture does the same to the natural powers of the soil. (See Selection VIII G.) He emphasized that there are two sources of wealth, nature and labor, and that progress in capitalist agriculture not only ruins the laborer but also robs the soil. (See Selection VIII D.) Marx and Engels saw the despoliation of plants, animals, soil, minerals, and other natural resources, but like their contemporaries they apparently were not alarmed by considerations of a limited and nonrenewable supply of such resources. They exemplified a general optimism in material progress—an optimism that originated in man's rise as an "ecological dominant" breaking into the accumulated wealth of various unexploited "ecological climaxes" on the planet in the past 1000 years and especially in the past 300 years.[33] For example, in *Utilitarianism*, first published in 1861, John Stuart Mill wrote: "Yet no one whose opinion deserves a moment's consideration can doubt that most of the great positive evils of the world are in themselves removable, and will, if human affairs continue to improve, be in the end reduced within narrow limits." Like most of their contemporaries, Marx and Engels did not realize that the rate of the depletion of these resources was asymptotic, that exploitable

frontiers were fast disappearing (the U.S. frontier was officially closed
in 1890, five years before Engels' death), and that the ultimate ecosystem
for mankind would soon be no longer a nation or colonial region but the
planet itself.

Marx compares the increase, misery, and capitalist exploitation of
surplus population to "the boundless reproduction of animals individually
weak and constantly hunted down."[34] He points out that as capital extorts
from the laborer the maximum of labor-power each day, without regard
to the length of life, so "a greedy farmer snatches increased produce from
the soil by robbing it of its fertility." (See Selection VIII I.) Thus, land,
animals, and men alike are "fair game" for the capitalist in his insatiable
search for prey. Such a condemnation of predation implies the importance
of the rational conservation of human and natural resources, a topic on
which Marx declared himself more than once. In a remarkable passage
showing an awareness of the reciprocal social and natural effects of an
ecology disrupted by capitalism, Marx states:

> Capitalist production by collecting the population in great
> centres, and causing an ever increasing preponderance of town
> population, on the one hand concentrates the historical motive
> power of society; on the other hand, it disturbs the circulation
> of matter between man and the soil, i. e., prevents the return
> to soil of its elements consumed by man in the form of food and
> clothing; it therefore violates the conditions necessary for lasting
> fertility of the soil. By this action it at the same time destroys
> the health of the town labourer and the intellectual life of the
> rural labourer. (See Selection VIII D.)

If ecology is concerned with the *interactions* of two or more living systems
in a natural environment, with the *interfaces* between systems,[35] then,
as we have noted, dialectics as "the science of universal interconnections"[36]
is a general science that embraces ecology.

Marx and Engels directed their studies principally to the dialectics of
society. Where dialectics touches on nature, it pertains frequently to the
"metabolism" or "exchange" of materials between man and nature, and
between the industrial city and the countryside, which contemporary
natural scientists such as Liebig had stressed. At this point, relying on contemporary chemistry, Marx and Engels were limited of course. They could

not be expected to know of the complicated cycles of natural elements, the influence of industrial processes on these cycles, the pyramids of life in ecosystems, food chains, and all the essential ecological facts uncovered in the twentieth century. During their lifetimes two great innovative thinkers on man's impact on the earth were doing their work—the American George Perkins Marsh (1801-1882) and the Russian Aleksandr Ivanovich Voeikov (1842-1914).[37] But Marx and Engels were not familiar with their work; they were humanistic naturalists (materialists), with the emphasis on *humanistic.* Yet they knew that if *human* is defined too narrowly and if man's vital interdependence with nature is neglected, the concept and the practice based on it become alienated and self-stultifying.

The philosophy of Marxism bases itself on man's need to breathe, eat, drink, and otherwise satisfy his material needs by feeding upon external nature before he can fulfill his higher, spiritual needs—for creative interpersonal exchange, aesthetic enjoyment of persons and nature, play, practical and theoretical understanding, imaginative symbolizing, artistic creation, and so forth. Marxism stresses the instruments and technologies by which such basic material feeding on nature is satisfied. It must therefore ultimately ask itself what are the limits to such feeding and technologies. In their critique of capitalism, Marx and Engels had already begun to ask that question. Yet they underlined in both theory and program the solution of socialized production, and we do not have from them explicit pronouncements on the problems that would face mankind at the end of the twentieth century. Even so, they recognized the general principle that the expansion of man's technological power to extract energy and goods from nature, changing and transforming it in accordance with man's needs, must be dialectically balanced by the maintenance of nature's fruitful integrity.

As Marx and Engels saw it, the overriding issue of their time was the exploitation of the great masses of industrial workers, and that was the point at which the consciousness of most men could be aroused. The primary places at which ecological damage was inflicted were the factory and the dwellings of the industrial workers, the large agricultural estates, and the rural slums; and it was to those places that Marx and Engels directed their main criticism. In *The Condition of the Working-Class in England in 1844,* Engels set forth the ecological dislocation produced among the industrial, mining, and agricultural proletariat by their urban and natural environments. In the *Grundrisse* and *Capital,* Marx expatiated on the effects of the machine in manufacture on the laborer. The laborer becomes a routinized, lifeless appendage to the machine, exhausted, unfree, deprived of interest, automatonic, unskilled, subjected to "a barrack discipline."

> Every organ of sense is injured in an equal degree by artificial elevation of the temperature, by the dust-laden atmosphere, by the deafening noise, not to mention danger to life and limb among the thickly crowded machinery, which, with the regularity of the seasons, issues its list of the killed and wounded in the industrial battle. (See Selection IX B.)

All the values, whether so-called economic values, like oxygen in the air or food, and so-called higher values, like music, have their ultimate origin in nonhuman nature. Oxygen is produced by the green plants; food is derived from plants and from animals feeding on plants; music comes from naturally occurring woods and metals and human bodies and the vibratory dispositions of these substances. These substances may undergo considerable transformation under the technical hand of man. The bio-social life of man feeds on such substances. It does not spring into being like an immaculately conceived virgin. Political economy is a humanizing, socializing, internalizing of these discovered natural substances and forces in the environment. It is a particular way in which a particular living species organizes its natural environment by locating, experimenting on, understanding, and transforming those substances. Thus, edible food, which has its source in living plant or animal organisms in nonhuman nature, has been deliberately cultivated in human society by means of long-developed techniques through cultural history. It is processed, transported, distributed, and eaten according to established cultural patterns. Human economy is interwoven with the living ecological relations of the society and is inseparable from those relations.

Hence, political economy when it is sufficiently generalized becomes ecology, and from time to time it was so generalized by Marx and Engels. For political economy pertains to the production and distribution of economic values and the social, governmental management of such values. At a deeper level it pertains to the basic laws that regulate individuals in human societies as well as one society in its relations to other societies, so far as these relations implicate economic values. *Oikonomia* is the management of a household—familial, communal, national. *Oikologos* is the study of households of organisms in general—the mutual relations of organisms and their environments, or bionomics. Thus, political economy can be seen as a species of the generic science of ecology; or we can reach ecology by tracing the most comprehensive, generic laws that underlie the transactions of human societies with their natural environments.

Marx and Engels aimed at the comprehensive view in political economy. They welcomed the method of Darwin in biology and had attempted to apply an evolutionary method (before they knew of Darwin) in political economy, penetrating beneath the particular surface forms to the underlying general dynamical laws. The logic of this method led them to investigate the economic relations of men and societies to their natural environments, and to pose questions about the optimal kind of man-nature relations. But they never fully developed the logic of this method, for two reasons: (1) the demands for a genuine science of political economy were so great as to call for concentration on the intrasocial relations and for a setting aside of intensive study of the man-nature relations, and (2) the demands for the relief of the masses of workers from starvation, sickness, wretched conditions of work and of housing, and the like were so imperative as to require concentration on exclusively "political" work and theory. It is true that Engels sketched some ideas on the dialectics of nature, but his thought on ecology was never very far developed. Yet an ecological position is there in the thought of Marx and Engels, embedded in the very method they employed in their theoretical-practical response to society and nature.

It is often said that the ecological crisis is somehow the peculiar problem and responsibility of the privileged plutocrats, the industrial barons, the oil tycoons, the great landowners, and the powerful politicians. It is not uncommon to say of such persons and of the crisis, as people have said of God after a heavy snowfall in city streets: "They put it there, let them take it away." This view is rooted in the misconception that a ruling class has ultimate power and responsibility. In his famous essay, "The Historical Roots of Our Ecologic Crisis," Lynn White, Jr., has pointed out that technology has traditionally been "lower-class, empirical, action-oriented," while science as a theoretical pursuit has been "aristocratic, speculative, intellectual in intent."[38] A reading of Greek philosophers like Plato and Aristotle reveals this class division. It may be added that the practical techniques of preurban man for a million years or more long antedated the speculations of theoretical science made possible by working classes and the leisure classes of ancient civilizations that commenced some 5500 years ago. Not until the middle of the nineteenth century did these two powerful forces become fused. Why?

White comments that this fusion "is surely related to the slightly prior and contemporary democratic revolutions which, by reducing social barriers, tended to assert a functional unity of brain and hand." But he leaves the answer in this very general form. The full and detailed answer was given in Marx's analysis of the system of capitalism. In the quest for expanding profits, capital employs and draws into its all-consuming system the knowledge of science and the methods of modern technology. To operate the machinery and develop the requisite ideas for understanding and exploiting nature's resources required the skills of both manual and mental workers. Both became cogs used indifferently in the productive process, and each group needed the other in order to make general ideas technically relevant to industrial production. The result was a tendency to integrate all workers, manual and mental, rural and urban, into the indifferent and all-powerful mechanism of capitalist production. While creating, without intention, this equality of working conditions, exploitation, and suffering, capital also engendered a demand for equal participation in the fruits of the collective process. Burdens and responsibilities were borne by the workers; corresponding rights were consequently called for. That demand under such conditions is a kind of human law. Hence, while the democratic revolutions did influence the democratizing of laboring men, the industrial process worked the other way in shaping political democracy. Collectively laboring in the modern factory, men came to realize that while they were creating all of the wealth of society, they were getting only a small part of that wealth in return. They demanded more, much more, of the surplus-value being stripped from them.

Marx and Engels sharply articulated this contradiction, its causes, and the path for overcoming it. Thus, in a strict sense the ecological crisis is not the creation of a few men alone. The much vaunted doom and promise of technology do not reside in the hands of these few alone. The crisis is the consequence of a system of relations between workers and owners, of a deepening contradiction between the forces of production (machines, techniques, workers) engaged in complex cooperation, and a system of ownership and distribution of the products that deprives those very workers of control over their products. Thus, a whole system, directed by the ruling classes and their hirelings in government, is involved in a self-destructive contradiction in which the workers are unwitting participants and victims. The system of production would be totally impossible without the workers and the tools and techniques of technology that they have created. Hence the contradiction and the consequent ecological crisis is their problem and it is their responsibility to solve.

Some thinkers suppose that technology and ecology are essentially *moral* problems. Jacob Bronowski, for example, writes that "technology has become a moral and not a material demand" on the part of the masses. He argues that making available to the masses what were once only upper-class luxuries—running and hot water, indoor flush toilets, health care and medicines, gas heating, electric lighting, and the like—is a consequence of the "moral rather than material indignation" of Karl Marx.[39] This argument misses Marx's whole point—that the moving forces transforming societies and history are the forces of production, among them the material drives of people to eat and clothe and house themselves and their children; and that moral demands and ideas are of no effect unless they ride on top of those forces and drive and guide them to their historical issue. We know, moreover, that moral indignation, organized in community groups, has little power to change things very deeply unless it is directed at the prime material causes of social injustice, namely, class ownership and control exercised in the interest of profit-maximization.

III. Capitalist Parasitic Ecology

Just as capital exploits labor, it also exploits nature, bringing upon society comparable disastrous consequences. Like labor (which, so far as it remains unconscious power, is in certain respects exploitable like the rest of nature), the world of nature is there to be used for the ends of capital, i.e., to transform its raw materials through labor so as to create, accumulate, and multiply use-values. The capitalist sees everything within the framework of his intention to turn a profit. Hence he views the things and processes as *gratis,* and they are taken as just that. They cost him nothing, but in order for him to obtain an increase of their value for his own purposes he requires the application of labor-power, some of which he must pay for but much of which he comes by free. Science, which is one form of productive forces, one form of "the social brain"[40] and which renders labor-power fabulously productive, is also free for the capitalist. (See Selection VIII A.) It provides the bridge from exploited labor to the fullsome exploitation of nature.

In exchange for its free gifts to the capitalist, what does nature require in turn? Nothing—at least nothing which it can vocally demand in a way that could be enforced in the courts or in human conscience.[41] For millennia, so far as it was driven to justify its appropriation of natural things and persons, the ruling class invoked

> the simple plan
> That they should take, who have the power,
> And they should keep who can.[42]

A man who works for a living, who tends plants or animals or handles clay or wood or metal or other materials in an effort to produce something useful for human needs, normally develops a healthy respect for the relations linking things and events. He comes to develop a working knowledge of the conditions, causes, and consequences of things. In contrast, a member of the leisure class, presented with the ready-made goods of life and rich and powerful enough to secure the fulfillment of many of his wishes, will be inclined to view the universe after the form of his own mental and social world. It will be for him a world of caprice, or arbitrary decision, of things easily given and easily taken away, of action and pleasure without perceivable cause or external consequence. His slightest wish will appear as a command to others and to nature.

Modern ecology and dialectics have confirmed the insights of ordinary workers throughout history and prehistory: things and events do not come out of nothing or pass into nothing but are caused, conditioned, transformed, and conserved in their consequences in a nature that is a self-enclosed system of energy exchanges. Nothing is isolated, nothing totally disappears, nothing is free;[43] everything is connected, everything has consequences, everything is the product of the workings of nature or the labors of man. Citing Rachel Carson's *Silent Spring* and her use of W. W. Jacobs' story, "The Monkey's Paw," Garrett Hardin has written:

> . . . *we can never do merely one thing.* Wishing to kill insects, we may put an end to the singing of birds. Wishing to "get there" faster, we insult our lungs with smog. Wishing to know

what is happening everywhere in the world at once, we create
an information overload against which the mind rebels, responding by a new and dangerous apathy.[44]

But he omitted the most momentous of foolish human wishes: wishing to
get rich, the capitalists have debased men and defiled nature. (True, "we,"
as the great generality of people who work for a living, are often heedless
partners in pollution. But the primary and most pervasive polluters are the
powerful corporate producers—the makers of insecticides, hydrocarbons,
automobiles, and advertisements.)

Only stupidity or blind arrogance would lead a slave-owner or a capitalist
to believe that nature was "given" just to him or his kind to do as he
pleased and like Midas to fancy that no consequences would follow or
that he could erase or flee them. Only incorrigible egocentricity would
permit him to think that only his own use and enjoyment of soil, air,
water, plants, and animals mattered, regardless of the consequences to
nature, his heirs, or other people. Yet it is just such arrogant attitudes
that have characterized the ruling classes throughout history and still
characterize them today.

In the works of Marx and Engels, there is a tension between the dynamic,
technological, industrial thrust of human history, usurping more and more
regions for the dominance of man, and the relatively stable laws of nature
that biologists describe as "the balance of nature." That is, the technological productivity of man toward nature is conceived to develop indefinitely;
but nature, as a limiting condition and "power in its own right," must
ultimately check that productivity. Ecology shows that in a common food
chain on the land the grass feeds off the sunlight, air, water, and organic
and inorganic matter in the soil, herbivores eat the plants, carnivores eat
the herbivores and die, and decomposers bring the original matter back
to the soil. A large increase in the number of men as carnivores places a
heavy strain on this food chain and is apt to disturb it at several points.
Moreover, the technologies of modern industrial societies, as a result of
the introduction of new quantities of elements into nature, have disrupted
its rhythms and endangered the life and well-being of man.

No one following in the spirit of Marx and Engels could sanction such
dehumanizing and denaturalizing results. Yet Marx and Engels could not
anticipate them, and it becomes the responsibility of Marxists to deal with
these new problems of the ecology of technology. New knowledge and
practices mean always, as Engels said, that materialism must "change its

form,"[45] and that, in Lenin's words, "a revision of the 'form' of Engels' materialism . . . is demanded by Marxism."[46] Marx and Engels laid down the basic outline and method of dialectical knowledge, but by its very definition such knowledge must be continuously informed and brought up to date, so that it can become relevant and useful with regard to the life-and-death issues that men face anew day after day.

While there is a tension between the laws of the development of any economic system and the laws of ecology, we believe there need not be an incompatibility, any more than there must be an incompatibility between dynamics and statics. An economic system is in fact a special and human kind of ecological system, as a workshop is a social relation of production. It was pointed out earlier that political economy sufficiently generalized becomes ecology, while specifically applied to human affairs ecology becomes economics. Moreover, an ecological system is a dialectical unity of both the motion and the equilibrium of forces. What Marx tried to show was the essential incompatibility of the economic system of capitalist production (like all forms of class production) with itself and with the system of nature. He described the inherent self-destructive instabilities of this system. Its "historical mission" is the "unconstrained development" of labor's productivity.[47] It is the unwitting agent of "a general social power"[48] whose "methods of production . . . drive towards unlimited extension of production, towards production as an end in itself, towards unconditional development of the social productivity of labour."[49] But this movement comes up against a limited and obstructive end, "to preserve the value of the existing capital and promote its self-expansion to the highest limit."[50] With the advance of skill and machinery, as the productive power of labor increases and reduces the number of laborers, the surplus-value increases, but as the total mass of labor is reduced, there is an immanent limit to this increase.[51] Such increased productivity is accompanied by an increase of commodities, an expansion of markets, an accumulation of capital, an overpopulation of laborers, and a falling rate of profit.[52] Labor and capital become concentrated, competition intensifies, and monopoly grows. The responses of capital are to render existing capital unproductive or to destroy it,[53] raise the intensity of exploitation of labor, depress wages, cheapen the elements of constant capital, produce surplus population, and extend foreign trade and stock capital.[54] But the rate of profit continues to fall, and crises deepen. The system is unstable and ultimately self-defeating.

The early theory of capitalism, the Ricardian cybernetic system, justified the starvation and other miseries of workers. In practice, the theory proved itself unstable because competition reduced the competitors in the direction of monopoly.[55] All post-Ricardian systems that proceed from the premise of *laissez-faire* are doomed to run into similar difficulties because the principle of competitiveness is self-eliminative. The only question is whether this tendency will move toward private monopoly or public ownership of the environment and its resources.

In ecological terms, the explanation for this instability of capitalism is that the interactions of individuals and groups within the system are determined by an exclusion principle.[56] The principle is: two groups in an environment of scarcity of need-satisfying things cannot coexist indefinitely. Under capitalist production, the factor that produces the division into two groups—the owners and the producers of productive power—is the possession of social power (labor-power) by the owners. The power is then impelled by the principle of the infinite development of capital, by the capitalist's devotion to profit[57] and his intolerance for the falling rate of profit,[58] by his "greed for the labour-time of others,"[59] by his interest in production for *capital* (i.e., himself) and not "for a constant expansion of the living process of the *society* of producers."[60] But the exclusion of the producers from the cooperative control of the productive forces and from the consumption of their products because of an egocentric organization of society is in process of being overcome by "an unconditional development of the productive forces of society" for society itself.[61]

We can speak of exclusion because of the incompatibility of two social processes—"producing use-values" and "creating surplus-value."[62] The productive power of cooperative labor, as congregated under capital, comes into conflict with exploitative profit. Cooperation as a principle of social organization wars with despotism.[63] Yet the control commanded by the capitalist is of necessity delegated to others in the large-scale cooperative enterprise,[64] so that his antagonistic control is undermined by the demands of cooperative organization. The "greed and lust for power," which have been the "levers of historical development"[65] in class societies, must and will be eventually displaced by the mutual aid of classless society.

If we describe the situation under capitalism in terms of a food chain, the producers of food, while not directly consumed in the flesh, are consumed piecemeal by the daily bite of surplus-value out of their labor-time, by starvation wages, and by no employment whatsoever. The struggle to get food for oneself and one's family, to resist being used by others in their struggles to get food or surplus-food, is unremitting. In primitive herd societies, men do not prey on one another, not yet being, as Marx said, individuated by processes of exchange. They collectively prey on plants and animals. In bourgeois society, however, the individual worker stands exposed, so that what confronts him—the property relations—under capital is "the *true common entity* which he seeks to devour and which devours him."[66] In the exchange between predators and prey, the predators return only a portion of what the producers pass along the chain. Thus the chain is not completed. In nature the predators temporarily wax fat, and the producers languish and die. But a time comes when the pathic conditions engendered by predation commence to affect the predators themselves. A shortage of producers or breaks in the chain of transmission of food creates famine, disease, social disorder, and, among some animals, cannibalism and violence.[67] (The principal killers of men in prehistory were famine and disease; to these the class societies of history added war.) The unique character of human social systems riddled with a parasitic social class is that the producers preyed upon, unlike the prey in other systems, are capable of *understanding* the conditions of their predation and *acting* to overthrow those conditions and to create a new system of social relations.

Reference to the producer-predator relation is not mere analogy. More than once Marx applied terms such as "predatory" to the capitalists. Indeed, he designated capital as a "vampire" and "parasite." "Capital is dead labor, that, vampire-like, only lives by sucking living labour, and lives the more, the more labour it sucks."[68] Is this a mere figure of speech? By no means. A parasite is an organism living in antagonistic relations to another organism, deriving its food (and possibly shelter and protection) from the host organism, consuming and not producing, taking from and not adding to the system of energy linking the two. Thus, while a parasite is living, it is a degenerate form of life, not "free-living" in the sense of generating an independent contribution to the maintenance of the system in which it survives and reproduces. It is true that all organisms feed on organic or inorganic matter external to them, but in contrast to non-parasitic types the parasitic organism, dependent on other organisms for food, is an unproductive member of an energy system. In exchange for the necessaries of life—the tissues or materials needed by the host and available in the host or its products—the parasitic organism returns little or no equivalent. (In *commensalism* the feeding organism is benefited while the host is indifferently affected; in *symbiosis* both benefit.) In short, the parasite is an unproductive being appropriating the living energy from other beings. To the extent that it takes from and contributes little or nothing to the energy cycle in which it participates, it becomes obstructive or disruptive of the functioning and continuance of that cycle.

By such a definition, which corresponds to the realities of the world of plants and animals, capital is a true parasite. The case is even stronger than in ordinary cases of parasitism, for human labor-power, "living labor," is, alongside independently existing nature, the source of *all* wealth, and capital is value (dead, accumulated labor) "which appropriates value-creating activity."[69]

> ... capital buys [labor] as living labour, as the general productive force of wealth; activity which increases wealth. It is clear, therefore, that the worker cannot become *rich* in this exchange, since, in exchange for his labour capacity as a fixed, available magnitude, he surrenders its *creative power,* like Esau his birthright for a mess of pottage. Rather, he necessarily impoverishes himself, as we shall see further on, because the creative power of his labour establishes itself as the power of capital, as an *alien power* confronting him.[70]

Thus we see the creations of labor, disguised in the form of capital, rising up to alienate, possess, and conquer their very creator. The worker finds himself oppressed by the noisy, dirty, unsanitary, unsafe, monotonous, fatiguing factory where he works, squeezed by rising prices and falling wages and inflation, tossed about by recession and depression, lured and deluded by ads, seduced by entertainment, and face to face with a world of nature increasingly polluted, abused, and made ugly. How strange that he, the creator of multiplying material riches, the heir of an abundant and beautiful nature, should be so poor! If pre-Marxist and anti-Marxist political economy is confused about precisely how this happens, so also capital and labor have been confused. It is certain that this murky twilight of the economic gods cannot be dispelled, that labor cannot be liberated from its chains in the workhouse of capital, until the contradiction is unraveled.

Marx revealed the secret and solution of the contradiction in classical political economy. According to the classical formulation, the value of a commodity is the result of human labor added to raw material, and the value of labor is determined (1) by labor-time—a vague concept—and (2) by the cost of production of the laborer himself (subsistence, maintenance, reproduction). A contradiction appears when the laborer works x hours but is paid for only a portion of the value he adds to the raw material during that period. Marx showed that in the exchange of labor-power for wages, the price of the laborer's commodity, the laborer is deprived of the full value of his productive labor-power or living labor. The convenient ambiguity in the classical formulation was that the "labor," which the laborer exchanges for wages and which is said to be valued and remunerated according to its necessary cost of production and reproduction, is not a block of time or a specific job to be done but *living labor itself in possession of labor-power.* This commodity of labor-power, as Engels said, is "a very peculiar commodity": it is "a value-creating force, the source of value, and, moreover, when properly treated, the source of more value than it possesses itself."[71] By definition, capital cannot begin to restore to creative labor the equivalent of its creative power, since capital is neither creative nor concerned about a fair exchange. Capital "appropriates," labor "surrenders" "this noble reproductive power."[72] What the capitalist seizes "legally" is not only the excess of value after his cost-price has been paid, a "surplus" of creative labor, but also the creator himself. Both the tree and its abundant fruits lie open to the exploitation of parasites. This is a tree of knowledge too, for all labor involves practical skill and advances in its range and efficiency as it is guided by theoretical knowledge. With typical guile, capital absorbs "the general productive forces of the social brain" in the form of fixed capital and especially in the form of machinery.[73] In the machine we find labor confronting its creation as an antagonist in the most intense and absurd form. The question of whether the environment of machines (under capital) will control man or whether man will control machines (without capital) is the question of whether man will control himself and the products of his creative labor.

The question of whether the masses of people will control machines in the interest of a tiny ruling class is also the question of who will determine man's relation to nature. For machines are the extensions of man's innate and developed bodily and mental power by acquired natural means into and over the external powers of nature. Tools and machines enable man to penetrate the inmost secrets of nature—those of human history, the cell, the gene, biological evolution, the atom, physical evolution, and ecological structure. The growing assertion today, that we must conserve and value nature, is evidence not only of an awareness of the corruptive consequences of an economy governed by motives of private profit, but also of a newly won leisure among increasing numbers of people whose humanized sensitivities in relation to their natural environment have been awakened. Machines, which are the materializations of man's brain and hand power, are nature's way of discovering itself to itself, the cumulative creations of countless generations of productive social labor in the species' efforts to wrest a living from nature and a satisfactory adjustment to it. As machines increase in efficiency, take over human functions, and save human labor, they transform man's former relation of life-and-death struggle against nature into a new relation, one of free time, of leisure, and of opportunity for the fulfillment of distinctively human needs.[74] For the first time in his long and arduous evolution, man can begin to relate himself to external nature in noncompetitive ways, in ways that promise to fulfill his needs for nonutilitarian knowledge, beauty, play, recreation, the observation and tending of plants, animals, and inorganic substances in their myriad forms. In such new relations, when persons are freed from the deprivations, cares, and competition of a scarcity economy, they can rediscover at a higher level their prehistoric sense of identity with nature. Man can, as Engels said, "feel" himself "to be one with nature," but in a way purified of the alienating fears and anxieties that were once mixed with his primitive feelings of dependence. But this freed and purified relation of man with nature can never be achieved under the rule of capital, which doesn't give a damn for people or nature.

The workers' wrenching free from the parasitism of capital is no more inevitable than that parasitism is unnatural. Marx asserts that nothing "natural" prevents the exploitation of one person's labor by another, and he takes note of the prevalence of cannibalism among men. (See Selection I E.) Indeed, what Marx and Engels offer us is a theory of the semicannibalism of class societies throughout history. One may protest that Greeks, Romans, Englishmen, and Americans were never headhunters or eaters of human flesh. That observation, however, does not expunge the fact that the social system of capitalism fed off the bodies of scores of millions of black slaves, who died in the hunt, transit, and captivity, and the fact that its wars fed off similar numbers of soldiers and civilians. In their descriptions of the cannibalism practiced on children in nineteenth-century English capitalism, Marx and Engels might have employed the savage satire of Swift's *A Modest Proposal*—but did not. Engels, however, reducing the logic of Malthus to its absurdity, cites a state program for the painless killing of the children of the poor, proposed by a pseudonymous "Marcus."[75] The parasitism of capital, Marx shows, is such that it sets parents against their children; in the miserable sweatshops, "the wretched half-starved parents think of nothing but getting as much as possible out of their children." (See Selection IX E.) Parents sold their children and abandoned their newborn by the millions to the elements or foundling homes. If children were not killed at birth, they were slowly crippled as they grew up. This is the ecology of capital toward the human species. If one believes this description is an exaggeration, he should have a look at the 10 million hungry in the United States and the billion or so starving or semi-starving, most of them in Third World countries where capitalism still rules through national or multinational corporations—and at the oppressive conditions of the workers and the child labor in those countries.[76] Hundreds of millions of people suffer from blindness (vitamin A deficiency), anemia (iron and folate deficiency), endemic goiter (iodine deficiency), brain damage (protein deficiency), and other disabling and fatal diseases—all as the consequence of an exploitative economic system whose *excess* of private profits and armaments requires such *deficiency* of human resources.

IV. Criticisms of Marxist Ecology and Answers Thereto

Marx, Engels, and Marxism generally have been criticized for certain alleged positions on ecological matters: (1) they have pitted man against nature; (2) they have anthropocentrically denied the values of external nature; (3) they have overstressed the conflicts in nature and have understressed its harmony; and (4) they have denied basic human values.

In this section, these criticisms are stated and are answered from a point of view grounded in the concepts expressed by Marx and Engels.

Marx and Marxism have been accused of pitting "man against nature,"[77] not only in the sense of dividing man dichotomously from external nature but also in the sense of declaring that nature is man's inveterate and mortal enemy against which he must daily strive in order to wrest a living and the leisure of freedom. Thus, it is held, Marx accentuated the opposition of man to nature rather than his unity with it.

In their fight against the extradeterminism of vulgar materialism and religious idealism, against the doctrines of fatalism and predestination, against resistance of reaction to social reconstruction, Marx and Engels repeatedly called attention to the unique power of men to think about their world and themselves, to act upon nature and social and political institutions and to change them—in short, to be free from the blind determination of events and classes and free for the fulfillment of their own human capacities. In doing so, Marx and Engels were bound to show how men are differentiated from other creatures, to describe and exalt the character of man's unique freedom—not freedom from the material world but freedom in and through a mastery of its workings. Thus man is distinguished from the animals by his power to *imagine* the outcome of his possible action and then to *realize* his purpose through action. (See Selection VI C.) The very act of production, which Engels contrasts with animals' collecting, is an act of foresight. (See Selection IV F.) It is this consciousness, exhibited in the production of his means of subsistence, which makes man distinctive.[78] It enables him to become a "conscious master of Nature" and to make his "own history." (See Selection IV E.) Thus, man is not "against" nature either in the sense of being totally discontinuous with it or in the sense of contravening its structures and laws with his own purposes. Man is immersed in the qualities, forms, and tendencies of nature. He is composed of natural elements and has evolved in dialectical relations with the natural environment. Therefore, whatever values he achieves must be achieved within the constraints and possibilities of his body and those natural relations.

Aside from their practical political concerns, Marx and Engels were influenced by the new nineteenth-century ideas advanced by Count de Buffon and Alexander von Humboldt and others who gave evidence for the role of man's active and far-reaching intervention in shaping the character and course of nature.[79] Such ideas, correlated with the rising empirical sciences and the Industrial Revolution, were directed against the seventeenth-century concept of nature as a divine design to which man must obediently adapt as a responsible steward. It was just such a philosophy of cosmic equilibrium that led captains of industry like Andrew Carnegie in the nineteenth century to argue that the divine design had designated them as the stewards of the fates of the laboring masses.[80] Engels accurately characterized this older view as "conservative":

Nature was not at all regarded as something that developed historically, that had a history in time; only extension in space was taken into account. . . . Natural science, at the outset revolutionary, was confronted by an out-and-out conservative nature.[81]

Marx and Engels shared the attitude toward nature held by contemporary men of industry and commerce and by the millions of settlers migrating to new lands to struggle with the hardships of the frontier. Whereas eighteenth-century Europeans, for example, viewed America as a utopian garden of abundance, freedom, and harmony, the nineteenth-century immigrants saw the wilderness as an obstacle to be conquered[82] and as a reservoir of potential wealth to be subdued and transformed by the labors of man.[83] Marx and Engels, of course, criticized the utopian socialists who proposed a retreat to a bucolic existence close to nature. In contrast to "utopias of calm felicity," Marx and Engels proposed what Frank E. Manuel has called an "open-ended," dynamic utopia in which both man and nature are subject to constant transformation.[84] In this philosophy, time, change, and progress are the key concepts, not space, stability, and conservation. (Marx speaks of how capital strives to "conquer the whole earth for its market," and at the same time "to annihilate this space with time, i.e., to reduce to a minimum the time spent in motion from one place to another.")[85] The influences of industry, science, and colonization have radically transformed the older philosophy of nature and man's relation to nature into a new philosophy whose implications we have not yet fully grasped.

The medieval period in Western civilization—from the decline of the Roman Empire and the rise of Christianity to the seventeenth century—was marked by a sense of the spiritual unity of man, nature, and God. As a microcosm, man was thought to reflect the macrocosm of nature, while a supernatural God ruled over both man and nature. All things in nature were held to be interconnected, an analogical reverberating to one another in universal sympathy, and by an analogical reading of their "signature," to the script of an invisible God.[86] Medieval humanism (1100-1320) emphasized within this world view the nobility of man and nature, indissolubly united in hierarchical and intelligible order.[87] (Marx and Engels on occasion contrasted this order with the disorder under capitalism.) Such a sense of the unity of man and nature, however, was primarily anthropocentric and spiritual; it was not yet confirmed by scientific observation, experiment, and mathematical theory, and was compromised by ultimate reliance on supernatural revelation. Yet, the medieval conviction in the rational order of nature paved the way for modern science.[88] The capitalist and scientific revolution brought forth a new concept of the relation to man to nature, in both theory and practice. The formula was Bacon's "knowledge is power" and Descartes' "maîtres et possesseurs de la nature." The attitude and practice of man's scientific mastery of nature was a sharp reaction against the attitude and practice of man's religious submission to nature and his emotional, romantic idolization of it. Instead of looking backward to an ideal dreamtime, outward to a conservative nature, and upward to a release from this life, capitalistic men looked forward in space and time to the active fulfillment of their secular dreams in the near future.

Dialectical materialism was a reaction to both sides of this antagonism—a rejection of religious, passive idealism as well as of scientific, exploitative materialism. It affirmed both nature and man, in dialectical opposition and unity. It affirmed both "harmony and collisions," both "cooperation" and "struggle." It denied a preestablished spiritual unity of nature and of society in favor of a unity yet to be created in socialism and beyond. It denied a ruthless "struggle for existence" in favor of a final class struggle that would eliminate all economic exploitation from human history and lay the basis for a truly human, cooperative society living in cooperative relations with the rest of nature. That would be a society of "real human freedom . . . in harmony with the established laws of Nature." (See Selection IV H.) In 1876, Engels wrote movingly of the unity of man and nature that is yet to be achieved:

> . . . after the mighty advances of natural science in the present
> century, we are more and more placed in a position where we
> can get to know, and hence to control, even the more remote natural
> consequences at least of our most ordinary productive activities.
> But the more this happens, the more will men once more not only
> feel, but also know, themselves to be one with nature, and thus
> the more impossible will become the senseless and anti-natural
> idea of a contradiction between mind and matter, man and nature,
> soul and body, such as arose in Europe after the decline of classic

antiquity and which obtained its highest elaboration in Christianity.[89] (See Selection VIII F.)

The word "feel" is important here, for Marxists are often accused of taking a coldblooded, positivistic attitude toward nature. From their youth, Marx and Engels expressed an esthetic attitude toward nature, though their pre-eminent economic, political, and polemical concerns in their maturity issued in an outlook and a language that pertained to the understanding and control of nature and society rather than to their enjoyment. In *The Economic and Philosophic Manuscripts of 1844,* Marx insisted on an aesthetic mode of experience as an essential dimension of human life and fulfillment. While Marx never returned to this theme to develop it, it is clear from both his early and late writings that this was part of the "humanism" which for him defined "the goal of human development"—for which all science and politics are means.[90]

In his early fecund period in Paris, Marx pointed out that plants, animals, stones, air, and light are things in our objective environment through which, by our productive interaction with them, we maintain our bodily existence. Even more, such things constitute an essential part of our spiritual (*menschlichen*) consciousness, not only as objects of theory but also as objects of natural science and art. Marx calls them man's "spiritual inorganic nature, his spiritual food" (*seine geistige unorganische Natur, geistige Lebensmittel*).[91] For they feed the materials for perceptions, feelings, and thoughts to nourish the "consciousness" of man; and man's affective responses to them, his feelings, are their objective counterpart and fulfillment—"truly *ontological* affirmations of essential being (of nature)." (See Selection X B.) Man is a being of impulses, each with its correlative need and fulfilling object in the external world. Hence, the objects of his needs—as stones, plants, animals, persons—are "indispensable to the manifestation and conformation of his essential powers," just as the sun and plant elicit and confirm one another's being.[92]

So far as I know, Marx never elaborated this pregnant notion in which he intimately and dialectically linked man's subjective mental life to the objective and inorganic world around him. Marx was striving toward an understanding of the unity of the so-called material world and the so-called spiritual world by means of the modes of human apprehension. His reliance on Feuerbachian categories does not diminish the import of this notion. How does the material world affect and transform man? Materially, by sustaining the form and working of his body; and spiritually, by creating and transforming his feelings and thoughts. But that world is not divided into "matter" and "spirit" any more than man's being is. Hence, that world (and man, who is an integral part of it) must be reconceived in such a way that the peculiar kind of interaction of human body and external environment which issues in human "consciousness" (feelings, concepts, memory, prediction, imagination, and the like) provides a *continuity* of object and subject, of "material" and "mental." Even in Marx's day, the sciences were in process of establishing the continuity of the inorganic and the organic: the organic is a peculiar way of creatively organizing the inorganic. Marx in the passages quoted was searching for the continuity of the nonhuman and the human, of the inorganic-organic and the mental. In these passages, he appears convinced of such a continuity; but it is stated only and is not demonstrated. The continuity of the quantitative and the qualitative—how "energy" is related to the felt qualities of feelings, sensa, imagery, etc.—is still an unsolved problem for both the natural sciences and philosophy. Yet, we all know intuitively that they are intimately connected, and on reflection we can see that to speak of "inanimate" objects such as stones and air and light as "material" while our feelings of them are "spiritual" is an unnatural dualism that bedevils our conceptual and linguistic way of framing our experience. We know these things directly, by feelingful thought, by thoughtful feeling. The response of our bodies' molecules to stones or light is in principle the same kind of response as the response of those molecules to plants, animals, or other human beings. Vibratory energy, with its own vector quality, is the linking force.

Thus Marx and Engels did not repudiate the principle of unity of man and nature, an idea that in the Western world can be found not only in the seventeenth century and the medieval period but also in the Renaissance, the Greeks, the Jews, and the Sumerians, among others. On the contrary, they affirmed such a unity in their materialism, but their affirmation was qualified by the recognition of the strife and antagonisms in nature and society. Although often for philosophical and polemical purposes Marx and Engels argued for the unique power of man's freedom—"man is the sole animal working his way out of the merely animal state"[93]—any cursory reading of their work will show that they were thoroughgoing materialists repudiating all traditional and contemporary dualisms. As for man's alleged "emnity" toward or "alienation" from nature, is it not evident that nature has never been so ordered that man need do nothing in order to survive, that he must exert his energy and ingenuity to make a living, to feed and clothe and shelter himself, and to ward off the elements, disease germs, and his predatory class enemies?

After Engels sold his partnership in the firm of Ermen and Engels, he wrote to his mother in the summer of 1869:

> Since yesterday I have been a different chap, and ten years younger. This morning, instead of going into the gloomy city, I walked for some hours in the fields in beautiful weather; and at my writing-table in a comfortably furnished room, where one can open the windows without blackening everything with smoke, with flowers in the window and a few trees in front of the house, work is very different from work in my gloomy room in the warehouse looking out on the yard of a public-house.[94]

William Liebknecht, who lived with the Marx family for a period of time, has recounted the Sunday walks of the family to Hampstead Heath from Dean Street and later from the north of London:

> . . . our favourite walks were on the meadows and hills between and beyond Hampstead and Highgate. Here flowers were sought, plants analyzed, which was a twofold treat for the city children, in whom the cold, surging, bellowing stone sea of the metropolis created a veritable hunger for green nature. . . . And now our meadow between Highgate and Hampstead transformed itself into an asphodel-meadow, and we wandered among the hyacinths as happy as the blessed heroes. . . . we looked proudly down on the world from our sweet-scented asphodel meadow— on the mighty, endless metropolis that is the world and extended before us immeasurable, wrapped in a nasty, mysterious cloak of fog.[95]

It was not that Marx and Engels did not love nature; it was that they loved both it and mankind. Loving them, they felt the need to help them rise out of their poverty and oppression. That is why they spent most of their time and thought in the smoky, "gloomy city" under the ugly mystery of the fog instead of on the heights of the fragrant asphodel meadow.

We must ask, finally, what are the possible, alternative, practical atti-

tudes with regard to man's relation to his natural environment, and what
kinds of human beings are produced as a result of their applications?
A supernaturalist philosophy which regards nature as unimportant or
illusory, which makes heaven man's destination, and which promotes in
man indifference or disdain toward nature will lead to an impoverished
personality, one whose sensory life, whose feelings and imagination, whose
cognitive and creative connections with his world, will remain barren and
undeveloped. Could such a person really be called "human"? And even if
human, is that the kind of model whom we want for ourselves and our
posterity? At the opposite pole, a person who plunges into the wilderness
without reservation, stripping himself of all the instruments of civilization
except the minimal ones required for survival, may discover a primitive
and peaceful harmony with his natural surroundings; but he will suffer
from want of contact, stimulation, and communication with that "civilized"
life of the *polis* that makes man distinctively human. Whatever he carries
into the wilderness—his language, his perceptions and conceptions, his
habits of work, his tools—will be the creations of hundreds of thousands
of years of human history before him. The million-year-old existence of
Paleolithic man is evidence that the physique of *Homo sapiens* can survive
without civilization. But is that what we want? Consider again the alternative of capitalist mechanical materialism, which would like to bulldoze
away grass, trees, forests, and mountains and bring such a denatured world
under a planetary astrodome of big business, all ordered and engineered
for a "growth" economy of increasing profits for the would-be masters
of the cosmos. Here also there is evidence that man can survive under such
extreme denatured conditions, as he has survived under precivilized conditions. The rural laborers, displaced to the cities, jobless and poor, have
painfully survived through the centuries; the slum-dwellers today survive,
as do the children in urban apartments who have never climbed a tree or
seen a wildflower or wild animal. But is that the kind of unsanitary, un-
seemly, sick, suffering, and brief survival we want and truly need? Here
one should be excoriating capitalism instead of Marxism for "opposing"
both nature and man.

    Under supernaturalism man's labor, his creative activity with nature,
is a necessary evil, or, considering the soul, not necessary at all. Under
primitivism, a minimal amount of labor on nature may be acceded to
and may even become enjoyable, but such activity is not regarded as generic

and as continuous with and transformable into the creativity of man's "spiritual" labors at a higher level. Under capitalism, as Adam Smith put it, labor is a "sacrifice," a "price" grudgingly paid for what man needs in his ceaseless combat with nature. Marx places his own view of labor—as "attractive work," as "the individual's self-realization"—in constast to the narrow, negative view of capital held by Adam Smith for whom labor is external, forced, and repulsive. Man *needs* labor, says Marx. He needs the interruption of tranquillity, the overcoming of obstacles, the objectification of himself. Only in that path lie his freedom and happiness. But it must be social labor, scientific and hence general, inherent in the productive process and "regulating all the forces of nature." (See Selection X C.) Where is the "opposition" here? If man as subject needs to labor in this creative way, then he no less needs nature as his eternal object. If man were to destroy nature, to bomb it into oblivion or to pollute it into uselessness, what would his science investigate and turn to human value, what would his art depict and celebrate? Experiments would have no point, poems would wither and die. It would be utter madness, total antihumanism, for Marxism to preach or practice a doctrine that would destroy the very ground of man's science and art. Hence any "opposition" to nature in this sense is by definition impossible in true Marxism. The only opposition of man to nature that is tenable in this philosophy is the opposition in the context of a dialectical relation of the two in mutual maintenance and enrichment. For Marx the new mode of production under capital, struggling to break free of the social relations of such production, creates a new system of labor and a new kind of person who corresponds to it—a person capable of responding to the new qualities of things, a person "as rich as possible in needs, because rich in qualities and relations," a person "many-sided." For the creation of such a new, enriched person, "exploration of all of nature"[96] is required. Such "exploration" is not blind opposition and destruction. It is dialectical transformation in which both nature and man are preserved, changed, and created anew.

A second criticism of Marx and Marxism is that they are excessively anthropocentric and underestimate or deny the values of the external world of nature. For example, Lynton Keith Caldwell says that "Marxism, committed to economic determinism as an explanation of human behavior, has tended to be inhospitable to ecological thinking in relation to natural resources or the human environment."[97] Soviet industrializing commissars,

he continues, "showed little more concern for the full range of values
that might be realized in the environment than did their capitalist counter-
parts."[98]

It is true that Marx and Engels picture nature in terms of its utility for
man—as an arsenal, larder, and tool house for man's productive labor. (See
Selections VI C; VII B.) But nature is there prior to man's labor and alongside of labor is a source of material wealth, of use-values.[99] It is also true
that these are use-values *for man.* Yet value for Marx and Engels is an
objective relation between man and man and nature in which production
and consumption take place, and in that relation nature as "the primary
source of all the instruments and objects of labor"[100] participates in the
value process. Moreover, Marx and Engels never for once presuppose a
sharp dichotomy between man and the rest of nature, for man is both an
evolved natural form in receptive organic relations with nature and an
active shaper of natural objects and creatures. Domestic animals and
plants, for example, are the results of "many generations" of "man's
superintendence." (See Selection VI C.)

One critic asserts that Marxists do not see nature "as an object of concern in itself. Rather it has value only in terms of man's economic needs."[101]
It is hard to know what could be meant by nature "in itself," either in a
Kantian sense or in the sense of a discrete reality entirely independent of
our cognition and action. Of course, we can conceive of nature as an ongoing system long before the human species arrived on the scene. But our
conception of nature's independent content and structure is necessarily
an extrapolation from the qualities and relations of our own experience.
Moreover, now that we have arrived on the scene of nature and participate
in it, nature does not exist in entire independence from our values and
actions. That is precisely why we have an ecological problem, namely, the
problem of how we ought to relate the human and the nonhuman parts of
nature. Every discriminable entity in nature appears to have a certain independence from all other entities, but ecology tells us that every entity
appears also to be interdependent with a large number of other entities.
So while we can treat entities and groups of entities as objects of concern
in themselves, they must be objects of qualified and not ultimate concern,
for they are not isolated and they are not ultimate. To treat nature as a
whole "as an object of concern in itself" independent of man is not possible
so long as man is a part of nature. It is realistic to express concern in action
for the life and fulfillment of plants and animals in the whole system of

nature and of human society. But it is unrealistic to believe that such concern can be expressed quite apart from concern about human beings. Some people in the ecological movement in the United States today seem to be unrealistic in this sense. They want man to minimize his life, simplify his needs, reduce population and consumption, return to the land, revere nature, and cultivate handicrafts. Friends of the Earth say, "True affluence is not *needing* anything." There is some wisdom in this critique of anthropocentric urban complexity and in the expressed human need for the nurture given to and received from nature. But it is unrealistic to forget that man's very life as well as his fulfillment depend on his knowledge and control of nature's substances and processes. Marxism rejects this unrealism, and in its view of the man-nature relation, it is inclined to emphasize man rather than the plants and animals. If the assumption of the criticism is that Marxism has this emphasis, the assumption is correct. And like many modern humanisms, Marxism has sometimes overemphasized man's place in nature.

The logic of those who understand the interdependence of the environment and man is to maintain and strengthen the social and natural relations in which we live and move and have our being, for to maintain the self is to maintain such relations. Furthermore, our economic needs form only a part of our complex of needs, though surely, as Marxists continue to remind the idealists, a basic part. We have already observed that while Marx in *The Economic and Philosophic Manuscripts of 1844* spoke of man's esthetic needs in relation to plants and animals, he said little about them later when he was spending so many years in his research in political economy. But he certainly had those needs and others in mind when he recurrently called for the "all-round development"[102] of all persons as the goal of history and class struggle.

Only occasionally does Marx refer to animals, and usually in an economic context. He points out, for example, that Descartes defined animals "as mere machines" and thus "saw with the eyes of the manufacturing period," while for the Middle Ages "animals were assistants to man."[103] Presumably when animals are displaced entirely by machines as instruments of production, and when food is synthesized chemically, animals will enjoy a freedom not enjoyed since their domestication for food and labor in Neolithic times, and man's attitude toward them will likewise change with man's new freedom. However, Marx and Engels were too busy advancing the socialist revolution to spend their time, as many utopians did, in speculating about the state of things for either men or animals in the New Jerusalem. On the basis of all the evidence in their writings taken together and in their principal emphases and spirit, it would be incorrect to infer that they had no concern for the organisms and objects of nature. Their total ecological concern is quite evident, and that we can suppose to include the creatures of nature compatible with man's fulfillment within the web of ecological relations.

For example, when Marx speaks of man's alienation from nature, he means specifically from human nature in the other person. "Stupid" and "one-sided" private (capitalist) property forces us into the relation of *having* toward things and persons, of *using* them. But when private property is transcended, in a nonalienated, truly human relation, the eye becomes a human eye because its object is a human object.

> The *senses* have therefore become directly in their practice
> *theoreticians.* They relate themselves to the *thing* for the
> sake of the thing, but the thing itself is an *objective human*
> relation to itself and to man, and vice versa. Need or enjoy-
> ment have consequently lost their *egotistical* nature, and
> nature has lost its mere *utility* by use becoming *human*
> use. (See Selection X B.)

The liberated human eye enjoys things differently from "the crude, non-human eye." The liberated human being perceives the qualities of con-summatory value not only in specifically human creations, such as per-sons, music, beauty of form, and plays, but also in nonhuman things such as minerals. Such humanized perception is a dialectic of human subject and object, human or nonhuman. The manner in which objects become man's "depends on the *nature of the objects* and on the nature of the *essential power* corresponding *to it.*" (See Selection X B.) Thus, it is a mistake to say that the objects of nonhuman nature are mere empty things, with no objective character or value-properties of their own. It is a mistake, too, to think that Marx conceived of nonhuman nature as mere neutral stuff, as fodder to be fed into the machine of man's mind or of industry in a one-way traffic that has no consequences for nature. On the contrary, such a view is the view of capital, and the ecological policies of capital prove it.

To understand why Marx and Engels said so little about animals, it is germane to point out that in nineteenth-century England millions of agricultural and industrial workers were treated like so many cattle and in fact often worse. Marx quotes one authority among many, George Ensor: "The Scottish grandees dispossessed families as they would grub up coppice-wood. . . . Man is bartered for a fleece or a carcase of mutton, nay, held cheaper."[104] The farmland of the peasants was embezzled into the pasturage of the wealthy few, and the deer in the park of the wealthy became "demurely domestic cattle, fat as London aldermen."[105] Rare indeed are the men who, seeing themselves and their families desperately hungry, would not trap or shoot the nearest edible wild animal available, or the nearest tame one in the confines of a plutocratic estate. Ecological considerations go to the winds when the human economy is one of scarcity and a war of each against all, as it was for the vast masses of the poor in nineteenth-century industrialized societies. (This is why ecological problems are compounded in class societies but are more likely to be solved in socialist societies.) Marx was not ignorant of such scarcity. In the slums of London, he and his family suffered privations in abundance—poverty, hunger, damp cold, sickness, and the deaths of four children.

Humanists like Jeremy Bentham in the eighteenth and nineteenth centuries pleaded for the rights of animals, and John Stuart Mill eloquently argued in his *Utilitarianism* that the greatest amount of happiness and the least amount of pain be secured not only to "all mankind" but also, "so far as the nature of things admits, to the whole sentient creation." Humane societies for the prevention of cruel and "inhuman" treatment of animals were a natural accompaniment to the movements for social reform. Humane societies and conservation groups, however, tend to arise among the wealthy classes and high-salaried or professional persons impelled by a variety of motives: a sense of "ownership" and identity with one's country, a desire to protect one's own private holdings, humanistic idealism, an elitist fear of popular or socialist control of resources, and a diversion and displacement of energy from the radical transformations demanded in society. (Some Nazis were fond of animals and believed in the conservation of nature.) Often the ruling and affluent classes expend great energy and time on the protection of humanized animals rather than on the welfare of brutalized children in home and factory or on adult workers reduced to the level of animals. Their concern for animals is a displacement of human concern, for their class position constricts the scope of their expressed human concern. Child labor and industrial dehumanization continue on a large scale down to our own day. In this century, Sarah N. Cleghorn in "Through the Needle's Eye" has exposed the exploitation

of the poor by the rich, the triumph of a playful class of nature-owning
adults over the working class of industry-imprisoned children:

> The golf links lie so near the mill
>   That almost every day
> The laboring children can look out
>   And see the men at play.

A society whose members treat one another sympathetically and rationally
will tend also to treat nonhuman creatures in a similar fashion. But neither
social nor natural humanization will be possible on the scale required until
a socialist transformation of society effects a change in the ownership of
industry and agriculture and, with that, an economy of abundance rather
than scarcity, popular ownership of nature, and an extended responsible
attitude toward nature's creatures. In a society and ecology of scarcity,
governed by mutual predation, neither persons nor nonhuman creatures
are safe: all are endangered by genocide and ecocide.

For Marx and Engels, the urgent problem of their time was starvation
and disease among the great majority of the population, not ecology. In
the eighteenth century, the mortal enemies of mankind, the great killing
diseases, had reached their peak in the world, attended by extensive hunger.[106] In the nineteenth century, Marx and Engels saw these destructive
tendencies competing with and partially countered by the constructive
forces of industrialism, technology, and science. Hence the right response,
they concluded, was not to mourn with Malthusian melancholy or to
romanticize with displaced sympathy for the animals—but to organize the
masses. Marx and Engels had the immense vision of what scientific labor,
"the social brain," could do to eradicate once and for all the age-old enemies
of the race—famine, pestilence, and war. For in the past these killers were
the consequence of man's ignorance; in modern times they have been
perpetuated by the class organization of society in the face of the abundance and health promised by science and the peace promised by a classless society.

In an effort to extend the general line of Marx's thought, Ernst Bloch
has proposed the notion of nature as incomplete, as the yet-unexplored
and "unknown Subject of natural processes."[107] Under socialism, he argues,
man's activity might become merged with and embody this Subject of
nature. Bloch has thus raised the question of Marx's and Engels' attitude

toward the independent value and creativity of nonhuman nature and toward the most fruitful relation between man and nonhuman nature in the process of production. Bloch is critical of the exploitative relation of bourgeois technology to external nature, its alternate "paternalism" toward nature and its "idolization" of it. He expresses the hope that in socialist society people can enjoy the autonomous qualities of nature without regard to "doing" anything to it for their own appropriative needs.

Alfred Schmidt has pointed out that Bloch's view of the merging of man's activity with nature's creativity, of the eventual identity of nature and man, is incompatible with the materialism of Marx. For Marx, subject and object are always distinct both in labor and knowledge.[108] It is the view of Marx and Engels, he says, that "nature in its laws is, so to speak, *present to us in its completeness.*"[109] While Marx was aware of the qualities and autonomous but slow changes in nonhuman nature, says Schmidt, "dominion over nature turns it into an object."[110] "Whenever Marx writes of the 'slumbering potentialities' of nature, he is always referring to the objective possibility, inherent in nature, of its transfer into definite use-values."[111] Schmidt is still more critical of Engels' undialectical reduction of history "to the special area of application of nature's general laws of motion and development"—in contrast to Marx's wish that historical laws should *"vanish"* through the rational action of men.[112]

What was the position of Marx and Engels in this matter? Most of their statements on man's relation to nature are descriptive and prescriptive in mode and informative and directive in function. They are intended to be objective statements about the stimulus-properties of things and relations as they are and might be. They aim at directing human action to the control of such things and relations with respect to human needs and purposes and the human values that are the corresponding fulfillments of those needs and purposes. Marx and Engels say little about the value of things independent of human needs and purposes. We have already indicated the driving motive behind this stress on the conquest of nature for the relief of man's estate: their reaction against the prevailing passive, conservative, utopian, and lyrical concepts of nature, and their enthusiasm for science and technology when controlled by and for the masses of the people.

Yet, the language and attitudes of Marx and Engels toward nature are not altogether unequivocal. In his early work, Marx wrote of the unity of man and nature through heterosexual relations, social relations in general (since our social relations are relations of natural bodies), social labor, sensuous appreciation of nature, and industry—in short, what he called the humanizing of nature and the naturalizing of man.[113] He envisaged communism as "the *genuine* resolution of the conflict between man and nature and between man and man."[114] We have already called attention to Engels' sensitivity to nature, his feeling for it, expressed in passages that are usually overlooked or misunderstood because of his long-time interest in formulating the laws of nature in an objective way, and because of the dogma that this kind of objectivity is inconsistent with poetic language or with the existence of values in nature.

While they did not work out the details and implications, the position of Marx and Engels on nature is consistently dialectical. Nature is neither dependent object nor independent subject. It is neither something to be unilaterally subjugated under the hand of man, nor something to be adored in resplendent transcendence over man. Nature is the whole of the working of things and events in their complex and shifting relations; the members of the planetwide species man are a part of nature, interacting and changing with it. Thus, a correct view of it and of man's place in it would do away with the traditional dualisms of man and nature, subject and object, value and fact, and the like. It would exhibit the intimate interdependencies of nonhuman and human nature and their mutual transformation. A careful formulation of this view would show that the species man through its technology is an important agent in the transformation of nature into what we called human history, thus raising to a new qualitative level the whole process of natural evolution. But it would not neglect the evolutionary ties and deep material dependencies of this species with regard to its natural environs and history.

At the very worst, the question of values in nature in the writings of Marx and Engels is an open question. In my own reading of these writings there is nothing that finally forecloses the issue, in spite of what defenders or attackers of Marx and Engels may say. The logic of man's dialectical relation to nonhuman nature does in fact lead to the conclusion that the ground of values, if not the values themselves, is prior to and independent of man's conscious intervention in and enjoyment of nonhuman nature. For in the process of sensing, acting upon, appropriating nonhuman objects, man secures objects that are not man-made but still have use-value, i.e., properties that are value-yielding. The dialectical process between human organisms and environment which is creative of human value is illustrative of a general pattern of organismic-environmental activity by which species sustain themselves and survive. But the adjustive pattern of organisms is the adjustive pattern of molecules complexly organized. Molecules bond, and atoms unite or do not unite according to their valences. Value is not an epiphenomenon added to fact; it is an inherent activity of matter. What apeman does in his social group and with his tools prefigures what *Homo sapiens* does, just as the valuing activity of apeman is prefigured in lower primates, in earlier mammals, and in single-celled organisms.

A third criticism of Marxism pertains to its philosophy of strife. Some critics not only have pointed out *ad hominem* "Marx's aggressive and domineering impulses,"[115] but they also have asserted that Marx and Engels overstressed "the bloody drama" of history and nature and "overlooked the role of co-operation and mutual aid."[116] Whatever Marx's impulses may have been—and there is ample testimony of his tenderness, humor, devotion, play, protectiveness, and other such impulses toward his family and friends[117]—it is not clear from these criticisms how these alleged impulses are related to his view of society and the world. It is true that a central feature of Marx's philosophy is the emphasis on the class struggle. However, Marx did not invent the ubiquitous fact of class struggle in human history, nor, as he himself observed, did he deserve any credit for discovering what bourgeois historians before him had discovered—the existence of classes and the struggle between them.[118] On the face of it, it would be absurd to find a self-proclaimed socialist who was not aware of the role of man's *social* origins, feelings, thoughts, and practices in history. The fact is that both Marx and Engels said a good deal about man's sociality. The point of describing the role of the class struggle in particular phases of history was to try to arouse the masses to the realization that class struggles could be abolished forever from history and that men could live in a truly human, social way with one another and the rest of nature.

In contrast to the dog-eat-dog individualism of capitalism, from the time of their early writings Marx and Engels demonstrated an awareness of the deep sociality of man and its decisive role in the evolution of man. Cooperation, or community of labor, which can be found at the beginning of human history, is natural to man.[119] Its productiveness has "a history embracing thousands of centuries," shaping nature into values usable by man. (See Selection I E.) This "social productive power of labor"[120] transforms man into a collective being, stimulates his spirits, and heightens his efficiency.[121] In its capitalist form, it increases mechanical force, extends its field of action, makes the field of production more efficient, puts masses to work, excites emulation, produces continuity and many-sidedness in the operations of work, and economizes the means of production.[122] *Laboring men* do these things—not capital. Whether it be two or more men at work on a portion of the physical environment in space-time, in a rice paddy field or on an assembly line, the definition of man is always an ecosystem where *men* provide the guiding force. This "working organism,"[123] as Marx calls it, expands and develops, and the great changes of history must be understood by tracing the dialectics of its movements, from the "gigantic structures" of ancient slave societies ("Asiatics, Egyptians, Etruscans")[124] to the huge and complex modern factory. Marx, however, distinguishes different modes of cooperation through history. Until modern times cooperation rested on "relations of dominion and servitude, principally on slavery."[125] Now cooperation arise in the nexus of commodity relations and profitable exploitation. The productive power of labor *appears* to belong to capital, just as cooperation *appears* to be specific to it.[126] But the means of production, though concentrated in a few hands, comprises a social power, organized into social labor, structured by social cooperation and division of labor, and combined with the natural sciences.[127] Such is the new system of society, a new ecosystem breaking away from the old with its umbilical tie to nature and moving in transition to an ecosystem beyond itself. Cooperation thus plays a necessary role in man's evolution, but it must be seen in the specific dialectic of man's struggle with nature and with antagonistic classes. The development of cooperative labor, for example, should not be connected with "mystical" ideas; nothing "natural" has prevented cannibalism—or, we may now add, genocidal wars and nuclear holocausts. (See Selection I E.)

The one-sided emphasis on "the struggle for existence" by Darwin and his adherents led Engels to comment that before Darwin these adherents stressed "the harmonious co-operative working of organic nature" such as the mutuality of plants and animals. He points out that in the absence of Malthusian pressures of overpopulation, new species may evolve throug increasing adaptation.[128] At the same time, Engels was suspicious of a "conservative nature,"[129] such as pictured by Linnaeus and Newton, wherein nature is fixed and immutable. The true ecological view must encompass both balance and strife: "The interaction of bodies in non-living nature includes both harmony and collisions, that of living bodies conscious and unconscious co-operation as well as conscious and unconscious struggle." (See Selection IV D.)

   In 1875, in response to P. L. Lavrov's request for his comments on Lavrov's article "Socialism and the Struggle for Existence," Engels wrote that "the social instinct was one of the most essential levers of the evolution of man from the ape." (See Selection IV F.) But just as he refused to take the Darwinian-capitalist mode[130] as definitive of the evolution of nature, so he declined to do the same for the social model. Engels suggests but does not develop an evolutionary view linking man closely to nature. He finds "a planned mode of action" wherever protoplasm acts (in today's language, teleonomy). (See Selection VIII F.) He speaks of a "social rudiment" in insect states, productive animals with tools (bees, beavers), and colonies of Hydrozoa.[131] He takes notice of the evolution of the sense organs along with that of the brain,[132] and he points out that "man himself is a product of Nature, which has been developed in and along with its environment." Hence "the products of the human brain, being in the last analysis also products of Nature, do not contradict the rest of Nature but are in correspondence with it." (See Selection IV G.) How else may we explain not only the cognitive but also the *esthetic* responsiveness of man to external nature, other than through this evolutionary adaptability? When Marx said that "the *forming* of the five senses is a labor of the entire history of the world down to the present" and "history itself is a *real* part of *natural history*—of nature developing into man" (see Selection X B)—he adumbrated an idea of modern evolutionary theory: that human sensitivities to external nature (the other person, plants, animals, and so forth) emerge as adaptive responses to the regularities and qualities of beauty (and ugliness) in the environment.[133] When Marx said that "the eye has become a *human* eye" in relation to the stimulus of an objective person, which in turn is an object "for man" (see Selection X B), he anticipated the more general idea that the evolution of color in the world has depended on the evolution of color-perceiving eyes.[134]

   But the principle of evolution through cooperation within and among species, which is so richly documented today, was not preeminent in the thought of Engels. It could not have been; for one thing, he did not have the pertinent knowledge now accessible. Dialectical materialism as formulated by Marx and Engels tends to stress, in both description and rhetoric, the conflicts and antagonisms in history and nature. To this degree, it reveals the limits of its culture and of the bourgeois science and philosophy against which it rebelled. A part of this emphasis may be laid to polemics; Engels tells Lavrov that in Germany, where "false sentimentality" and idealism still rule, "it is hatred rather than love that is needed." (See Selection IV F.) A part of it is the lack of specific, ecological, evolutionary knowledge. To the extent that dialectical materialism lacked such knowledge, it was a philosophy of transition, moving from a feudal and capitalist view of nature and man to a socialist one but retaining remnants of a view that overstresses man's "mastery of nature."

Marx frequently makes use of figures of bodily, mammalian creativity in his works, such as "womb," "pregnant," and "birthpangs."[135] This concept of man's creativity plays a major role in his system of thought and helps elucidate his concept of nature. History, he says, is "the creation of man through human labor"; it is "proof of his *birth* through himself, of the *process of his creation.*" History is "the emergence of nature for man." (See Selection X B.) Thus, man's power to create his own mode of life, to wrench himself free from the constricting environment of nature, to cut the "umbilical cord" that ties him to both nature and his fellow man in an animal-like relation (see Selections VI H; VI I)—in a word, to stand free from nature's constraints and, within the limits of natural laws, to pursue his own independent and now self-imagined and self-fashioned development—this is what makes him truly human, truly noble. Thus, being born as a separate and distinct entity among the other parts of nature, man the self-creator can then begin to act back upon it and introduce a new factor in the whole evolutionary process.

That factor is the very creativity of his labor, which is cumulative and exponential both in what it produces and in its capacity to produce. Man's creativity is negentropic. It defies the blind drift of events and the clash and dispersal of their interacting energies. Man is the Promethean creator who steals the fire of nature, brings it down from the abode of the capricious volcanic gods, and ingeniously adapts it to the purposes of warming the bodies of men and lighting up their souls. Thus, the life of man transcends to this degree the closed system of prior inorganic nature with its physical "zero-sum" structure of simple action-reaction and cause-effect. But man's transcendence does not carry him beyond the natural universe. Its multiple-loop feedback system of dialectical theory and practice does not violate the physical laws of nature; rather, it differentiates, integrates, and refines them at a new level. His "birth" or "creation"—

i.e., a newly integrated structure of qualities, meanings, and activities—individuates him and thus marks him off from other individual things and persons in the world. So far as his creativity is liberating and humanizing, it separates him from dependence on persons and conditions that dominate him to his detriment—from ruling classes, private property relations, poverty, racism, sexist chauvinism, and political tyranny. At the same time, his creativity affirms his dependency on persons and conditions requisite to his fulfillment—his communication, cooperation, and solidarity with others in struggle for liberation, and his essential organic relations to the planetary ecological system.

A careless reading of Marx might lead to the conclusion that he believed that this creativity of man is wrested out of the world, as Hercules wrested the Nemean lion from its cave and strangled it, without any help from the world. Sometimes Marx in his passion and enthusiasm writes in this way. For example, he states, "A *being* only considers himself independent when he stands on his own feet; and he only stands on his own feet when he owes his *existence* to himself. A man who lives by the grace of another regards himself as a dependent being." (See Selection X B.) Yet the context shows that here Marx means to reject the domination of those who own the creative labor of a man, and not his interdependence with his co-workers or his relations to "his inorganic body, nature."[136] Man is a natural being, and like plants and animals he is destined by nature to bear as a patient a relation of need toward the things and persons of the world. In such a role of recipient, he awaits and yearns for the external things complementary to his need. (See Selection III B.) Such things are "indispensable to the manifestation of his essential powers." Such things are food for the hunger need, someone lovable for the need for love, something beautiful for the need for beauty. He cannot live in a human way without them.

Although Engels also recognized the dependence of man on natural things, he too was prone to make pronouncements that exalted man's power over nature and man's difference from nature. He wrote to Lavrov that he agreed with the substance of Lavrov's view that "the idea of solidarity could finally. . . . grow to a point where it will embrace all mankind and oppose it, as a society of brothers living in solidarity, to the rest of the world—the world of minerals, plants, and animals." (See Selection IV F.) In the light of Engels' frequent utterances on dialectics, we may presume that when he says "oppose" he also means "unite with," though the emphasis here is characteristically on the opposition and not the unity.

In a relation to other persons, in a relation of "solidarity," a person may find a mutuality of reinforcement, and hence the negentropic power of human creativity increases. The same cumulative relation may be developed between man and his natural environment through its attentive cultivation by man in accordance with the laws of conservation and growth But eventually the biological being of each individual man, composed of the elemental particles, decomposes.[137] Death occurs, and his biological individuality returns to the less differentiated continuum of inorganic nature. At the same time, the psychic life of the human individual dies, though some of his creations are preserved in the external material world or in symbolic form in living persons. Marx therefore could agree with Shakespeare on the enduring quality of the shared creative life of men, on the capacity of one man's labor to embody itself in some form of symbols and to live on in others after the bodily death of the creator— with the important material qualification: "so long as men can breathe and eyes can see." For the riches of human living are to be found not in material wealth, power, and fame, nor in the satiety or adornment of the body; they are cultivated, consummated, and enjoyed in man's creative, spiritual life:

> Within be fed, without be rich no more:
> So shalt thou feed on Death, that feeds on men,
> And Death once dead, there's no more dying then.[138]

In Marx's own youthful and Hegelian-Feuerbachian terms, the determinate species-consciousness of the individual confirms his social life, and *vice versa,* and the unique individual can attain the thought of universality of his species. While death may seem "a harsh victory" of species over individual, particularity is mortal while the species carries on. (See Selection X B.) Thus the enduring, evolving unity is the species—or, more correctly, as modern biology has demonstrated, the total system of living things and its environment. This viewpoint links man to nature's powers and destiny, as Engels realized when he wrote that matter will exterminate man on earth, "the thinking mind," but "must somewhere else and at another time produce it."[139]

For Marx and Engels man's body is composed of the elements of nature; man depends for his life on the air and water and earth, he is conscious of its qualities, and he forms and shapes its materials to his use. But they were disposed to stress his independence from it rather than his dependence upon it. The reason was simple: people were needy, work had to be done, and dependence of this elementary kind could be assumed. One contemporary scientist describes man as "umbilical to earth."[140] But for Marx and Engels as champions of the Enlightenment, such a romantic affirmation of man's primitive ties misses the demand of reason. Only reason, they thought, could rescue man from the wretchedness of class society. The "birth" of man and the severance of his umbilical cord with nature do not mean that man ceases to be an animal, a constituent part of the system of nature. They mean that man emerges from the blindness of instinct and detaches himself not from his organic metabolism with his environment but from his helpless subjection to its laws in nature and society. With the emergence of reason, man integrates conscious foresight with instinctual needs. Through such a route, hunger becomes socialized appetite, lust becomes love, protective defense becomes tender care, and the objects of nature, previously regarded with anxiety or propitiation, can be used, cooperated with, humanized, or, if malign, neutralized.

As man's needs are tied in a steadfast integrity to nature, there is no way to alter and fulfill man without also fulfilling nature. Marx and Engels talked about altering man in profound ways, and they assumed (but seldom elucidated) that nature would be altered correspondingly. Thus the primitive remains in man, deep and enduring. That is never denied. That means that he shares similar needs toward nature with primitive peoples, other mammals, primates, and many more creatures. The problem of man's development is not to deny those needs but to humanize them. This task requires a more intimate and not a more distant relation of man with nature. An ironic accusation made of Marxism is that it is both "animalistic" in its view of man and neglectful in its view of nature. But if man is a human animal, as Marxism maintains, man's needs require the full support of both society and nature.

In one of his treatments of this problem, where needs and their perversion under capitalism are being discussed, Marx points out that the modern worker lives in "a cave dwelling . . . contaminated with the pestilential breath of civilization." Thus "even the need for fresh air ceases for the worker." More precisely, we can say that the need is there and remains there so long as the worker strives toward optimal health; but it has been suppressed and alienated by the suppression and alienation of the environment. The needs for light and air—"the simplest *animal* cleanliness"—are met by "filth," by "stagnation and putrefaction," by *"sewage."* Hence, in such a world, man's *senses* deteriorate to a level even below that of animals. Sensory need is therefore virtually extinguished. "The savage and the animal have at least the need to hunt, to roam, etc.—the need of companionship." (See Selection IX F.) These needs are evoked and ful-filled in a healthy natural environment, but the natural environment under the control of capital is niggardly and polluted. Capital scorns to make it livable even in a minimal sense because it scorns the life of the workers. The problem of ecology under capitalism can thus be summed up as follows: capitalism pollutes and destroys nature because it pollutes the lives of the workers, who need nature to live healthily. Capital frag-ments and corrupts man's unity with nature; socialism restores it. The pollution of capital is total. That is why the cleaning out of capital must be total.

A fourth criticism pertains to Marxism's alleged lack of humanism. Since its beginnings communism has been reproached as desiring to do away with all the values claimed to be essential to human life by the critics. In the *Manifesto* Marx and Engels mentioned some of the values which their critics reproached them for wanting to destroy: the personal acquisition of property, independence, freedom, the work ethic, the family, the protection of children, home education, the rights of women, countries and nationality, "eternal truths" such as Freedom and Justice, and "all religion, and all morality." A reading of the *Manifesto* will reveal the ignor-ance, confusion, hypocrisy, and error in these charges as made by defenders of bourgeois society. Unfortunately, in 1847 ecological concern had not yet developed to the point where critics thought to accuse communists of dereliction in this domain too, and so Marx and Engels did not then have the occasion to refute this slander too. But the demand in the *Manifesto* for "the bringing into cultivation of waste lands, and the improve-ment of the soil in general in accordance with a common plan" is a clear declaration on the collective care of the earth. Yet these criticisms are not insignificant, for they have been elaborated and repeated for more than a century, and they attack the alleged antihumanism of a Marxism which is originally and deeply humanistic in goal and spirit.

A recent cold war spokesman, Fred Schwarz, has put this criticism anew: "Building on the doctrines of godless materialism, Communism has completely reversed the meaning of our basic moral terms."[141] He claims that human (Christian) values like peace, truth, righteousness, and love have been perverted by the communists. At a more fully philosophical level, Marx's theory of alienation, the essential dialectical negative aspect of his theory of humanism, has been derogated by sundry detractors. Until the postwar translations of Marx's early writings in Western Europe and the United States, anti-Marxist critics, denying the pervasive though not fully explicit humanist strain in the work of Marx and Engels, claimed that Marx and Engels were "economic" and "materialist" and hence antispiritual (antihumanist). When the translation of works like *The Economic and Philosophic Manuscripts of 1844* appeared, with a clear theory of the alienation of man's personality under capitalism, as well as an articulated humanism, a new tactic of the anticommunists had to be invented. If the impact of Marx's materialist humanism could not now be denied, somehow it had to be blunted.

Some philosophical critics, sympathetic to socialism if not communism, responded by drawing an implicit if not explicit distinction between Marx's materialism and his psychological theory (or "spiritualism"), between his theory of class struggle and his theory of alienation-humanism.[142] Jean-Paul Sartre, for instance, argued that while Marxism correctly dealt with the problems of society, only existentialism could explain the phenomena of subjectivity, consciousness, and freedom.[143] The anticommunist Robert Tucker maintained that Marx was mistaken in finding man's self-alienation to be "the essence of capitalism." Instead

> alienation only becomes in a derivative way an economic fact.
> . . . Marx originally had within his grasp, or at any rate within
> easy reach, the truth that alienation is essentially a fact of the
> life of the individual human self. . . . But he failed to trace this
> egoism to its real source within the alienated individual himself.[144]

Is the source of hunger in the "individual himself"? Of disease, bodily or mental? Of joblessness? Of war? Of the energy crisis? Of pollution and other ecological problems? Are we to believe what was told to Job's tormentors more than 2000 years ago, that these things are the result of individual sin?

Louis Dupré, while praising the "profound humanism" of Marx, says that Marx "grossly oversimplifies" human praxis "by identifying it with a fulfillment of physical needs."[145] Yet such a submergence of praxis in nature, according to Dupré, empties consciousness of content, and the dialectic of man with nature is thus done away with.[146] Furthermore, "without alienation there can be no dialectics," and in a nonalienated communist society, therefore, there would be no dialectic.[147]

"But even more unfortunate, from a philosophical point of view, is Marx's reduction of man's alienation to a social-economic level."[148] So the conclusion is that, while Marx's humanism was "profound," it was insufficiently sublime; while it was immanent in the material world, it was insufficiently "spiritual" and transcendent. The consequence is that its rigidity in systematic theory "as a rule of action leads to terror and constitutes a dangerous threat to human dignity."[149] Hence, in such an interpretation in which the answer is given *a priori*, Marxism *without* humanism forty years ago led to terror, while today the same philosophy, now *with* humanism, also leads to terror. Soon we will be reading that precisely because Marxism has a sensible ecological position (recently discovered by its critics), it is polluting the environment.

By reference to Marx's own texts, especially *The Economic and Philosophic Manuscripts of 1844,* I have already tried to show that for Marx human praxis is *not* identical with "a fulfillment of physical needs"—if by "physical needs" is meant survival needs. Marx's view does *not* submerge praxis in nature, *nor* empty consciousness of content, *nor* do away with the dialectic. In man's individual and social transactions with the environment, man satisfies his needs to eat, drink, clothe himself, and otherwise survive; he also fulfills his "higher" needs for beauty, love, knowledge, creativity, and the like. Of course, for Marx this whole hierarchy of needs is psychosomatic from beginning to end, since Marx repudiates all dualism in man and nature and presupposes the principle of emergent levels of events which are enveloped in and dependent on prior levels. If by "physical" Dupré means what is limited to the psychosomatic organism of man and the natural environment, and if by "spiritual" he means what is ontologically *beyond* the body in this sense, then of course he is correct in his characterization of Marx's position. Marx is not a supernaturalist. But there is no rule in ordinary usage or philosophical reason that compels us to employ the term "spiritual" in a supernaturalist sense. Nor is it self-evident that supernaturalism is the correct ontology and anthropology. For reasons of logic and experience, materialism or naturalism seems to Marx a more adequate world view than supernaturalism, though this is not the place to set forth such reasons. In any case, dialectical materialism does not truncate the spiritual or banish it from human living.

Let us turn to Dupré's charge about the concept of alienation in Marx. The great dramatic theme of Marx's life work was the theme of community-alienation-community. He viewed man's evolution as moving from pre-historic, primitive, unindividualized communism into the estrangements of class society and history, and thence into the eventual supersession of this estrangement by means of the new human attachments in communist society. In communist society human beings will come into at-homeness with themselves and their world by associating in free, fulfilling ways with one another and with nature. The broken bonds of man and man and man and nature will be restored. In the manuscripts of his youthful Paris period (1843-1847), especially *The Economic and Philosophic Manuscripts of 1844,* Marx sketched the outlines of man's estrangement from himself, his work, others, and nature. In the massive, sprawling, coruscating manuscript of the *Grundrisse* (1857-1858) Marx focused on the theme of alienation in economic life. He crystallized the misty Hegelian categories of his youthful thought into the hard facts of capitalist produc-tion as exposed by empirical investigation. *Capital,* first published in 1867, brought the results of this economic research into more precise and sys-tematic form.

It is a fact too frequently overlooked that in *Capital,* often set aside by half-hearted readers as "dull" and "dry," Marx quotes extensively from the heart-rending reports of government factory inspectors, the Children's Employment Commission, and others who in graphic language disclose the same sordid conditions in industry as pictured by Charles Dickens (whose works Marx read with appreciation). For example, in one cottage industry in England where pillow lacemaking was carried on, children as young as two years of age were employed, the older children working as long as sixteen hours each day, in cold, overcrowded cottages of as many as eighteen persons, surrounded by "drains, privies, decompos-ing substances, and other filth . . . where the smell was unbearable." In the straw-plaiting and straw-hat making industries, children in their third year were crammed into spaces sometimes less than 3' x 3' x 1½'. "The straw cut their mouths, with which they constantly moisten it, and their fingers." (See Selection IX E.) It was such conditions of degradation in the very core of human life, the child in the family, that led the young

Marx and Engels to cry out: "Do you charge us with wanting to stop the exploitation of children by their parents? To this crime we plead guilty."[150]

The repeated references to children in the massive tome of *Capital* are not mere humanitarian embroidery on a "scientific" treatise. They are essential to the argument that the capitalist system is "blood-sucking," that it inhumanly appropriates human labor-power from the cradle to the grave, and that if their stolen treasury of unrealized human talent is to be regained by the people such a system must be overthrown by the people. Marx recognized that the child is the generic human being in all its potential, the initial stage and promised fulfillment of human maturity. He saw that to damage the child is to damage the genus of the race, the very image of humanity, the bud of the possible blossom and fruit. Destroy that bud and you destroy the blossom of youth and the harvest of mature years. Starve, confine, expose, and beat that young fragile body and you cripple the free and full development of the body and the unfolding of the soul.[151]

In a remarkable passage anticipating some present-day psychological studies, Marx refers to the continuity of childhood and adulthood as well as the qualitative difference between the two. Discussing the charm of Greek art, he says: "A man cannot become a child again unless he becomes childish. But does he not enjoy the artless ways of the child and must he not strive to reproduce its truth on a higher plane?"[152] One of the profoundest alienations produced by class society is the separation between the uncorrupted human tendencies of the child—for perception, feeling, empathy, cooperation, knowledge, creativity —and the physiologically grown adult who has realized very little of such possibilities.[153] (A parallel alienation exists in thought between the present-day generation and the childhood period of the race.) Marx had the vision of an unbroken, all-round development of all persons from birth to death. Each person must be continuously realizing his human (generic) and individual possibilities through the full trajectory of his life, "reproducing" his potential self on successively higher planes as he passes through the temporal phases demarcated as childhood, youth, middle age, and old age.

Early in his student career, Marx turned from the humanistic idealism of Hegel's philosophy to a form of humanistic naturalism. In his Notes to the doctoral dissertation, *Differenz der demokritischen und epikuraischen Naturphilosophie* (1839-1841), he stated: "The point from which the old

Ionian philosophy proceeds on the basis of its own principle—seeing the divine, the Idea, in nature—is the point to which modern rational perception of nature for the first time must rise."[154] But having perceived man as the only divine being, and as integral to nature, he soon came to perceive also that the "alienation" of which Hegel had written in glowing, cosmic, and spiritual terms was a quite material, observable process of alienated labor. Such labor within capitalism alienates man from the rest of nature. Historically this alienation is a feature of an economic development having two opposite sides, closely linked: the separation of man from the domination of natural forces, from "animal behavior toward nature" (see Selection VI I), from the "umbilical cord"[155] and "leading strings" which hold primitive man in thrall to blind tendencies hitherto dominating man; and, later, the separation that arises at the technologically more advanced level of class society, wherein man explores, discovers, and manipulates natural forces to his own advantage but simultaneously misuses and misunderstands them to his own suffering and peril.

Marx concentrated his analysis, however, on the estrangements rending society, the antagonisms of labor and capital, of class and class, of rich and poor, of person and person, of family member and family member. People undergo the psychic consequences of such antagonisms; anxiety, anger, and grief possess them within the separated body of society. They protest the sundering of fundamental human relations and the healthy relations of persons to nature, and they strive to overcome such separations. Individually and collectively, people demand, first gropingly and then clearly, the restoration of attachment, cooperation, and unity with one another and with nature. This demand is the mature and social counterpart of the infant's anxiety, rebellion, grief, and struggle against separation from his mother.[156] When energized, organized, and informed, it takes the course of social revolution against the basic conditions and causes of these separations.

Socialist society on a national and international scale will bear witness to the recovery of man's elementary relations of interdependence with nature and other human beings, though at a level higher than that experienced in prehistory and history, a level more sensitive, more conscious, more humanly directed, more reliant on the material and spiritual resources of nature. The maternal principle of Neolithic "Mother Nature" will be taken up into a new, more intimate, more cognizant relation to nature. The principle of authority which has hitherto been monopolized and obscured by kings, priests, recent class rulers, and other father-figures will be absorbed into genuinely free and democratic rule. The older naturalism and humanism will be recognizable only so far as those primitive, ancestral forms survive not as doctrines but as transformed features and aspects of a new unitary perspective—a planetary outlook and way of life in which human beings are so fully naturalized and nature is so fully human ized that it no longer makes sense to speak of "man" and "nature" in separate terms.

Thus, socialist society will embody the culminating stage in man's relatedness to other persons and to nature. In primitive society, man has a dumb, inarticulate, uncomprehending relation to nature. To him nature is a void, mingling with man's void, a great unbounded unknown that is shot through with sunshine and starlight, with thunderings and lightnings, with fires and floods, with events benign and fearsome. Society is also seen as a nebulous herd, a mass from which individuals have not yet distinctly emerged. If, as in Neolithic or early civilized times, rulers have emerged, they are conceived to bear some vague, magical relation to this mysterious cosmos which endows them with the secrets of its power. At an advanced stage of class society, in which science and technology have sharply demarcated and ordered the events of nature, nature is perceived as an object of utility, as a set of discriminable individuals to be manipulated and used. Nature has emerged from the mists of the primeval religious imagination to become a secular Other which (as land) man might appropriate, (as animals) he might tame and harness, (as water) he might transform into power, (as space) he might conquer. Natural beings now seen as potentialities and no longer as finished things, now seen as manageable energies and no longer as alien forces, arbitrary boons, or threats, become a field for exploitation. So do human beings. Nature is energy, energy is work, people are workers, whether slaves, serfs, or wage-laborers. Such is the age of utility, in which ruling classes use both nature and men as instruments to their own power and pleasure.

Under socialism this is all changed. Human need, which in primitive times lived constantly in or on the edge of scarcity, discovered that by aggressive transactions with nature it might be fruitful and multiply. In class society, in the war of each against all to satisfy human need, the result was social chaos and irrational extremes of wealth and poverty. From a mild, passive, and propitiating thing, need became active, desperate, manipulating, and manipulable. The promise of abundance under technology expanded and encouraged human need; the fact of poverty contracted and discouraged it. Under socialism there is and will be abundance socially ordered for all essential human needs. Moreover, human beings in their relations to one another will no longer be dumb or voracious, will no longer be forced to treat one another as mere portions of the environment or as mere means to their ends; they will treat one another as ends in themselves. In Marx's words, where man's object is a human object, "need or enjoyment have consequently lost their *egoistical* nature, and nature has lost its mere *utility* by use becoming *human* use." (See Selection X B.) And what of nonhuman nature, the animals and plants and rocks and water and air? We know from observation of persons whose essential human needs are secured, who are therefore not egocentric but oriented to the welfare of the human species, that nature as such tends to be respected. With human needs satisfied and made safe from future deprivation, there is no need to exploit natural beings, and so exploitation does not occur. So far as possible, then, natural beings can be studied, learned from, enjoyed, and sometimes befriended.

This succession of stages has imperfect parallels with the dialectic of Master and Slave in Hegel's account, to which Marx tried to give material and historical meaning. At first primitive man lives in a relation of unfrustrated, or at least unconscious, desire toward nature. (Eden is in the tropics, where man originated; it is ecological bliss.) But at some point— in class society, if not before—the drive of desire comes up against an adverse world of nature that no longer furnishes what is needed for survival. Nature is then transformed into an Other, negative and estranged, as are other persons and classes who oppose the upsurge of desire to fulfill itself and to live. In this relation nature dominates man and subjugates him to its apparently despotic will. It enslaves him, forcing him to live by the sweat of his brow and to extract his sustenance from a nature that does not readily nurture him. But in the very demands of this dialectical relation man as laborer, responding to his own desirous need and the apparent scarcity of nature, adds a new dimension to the relation. Now *he* begins to dominate nature. He learns its laws, he applies his powers to it, and he extracts his sustenance from it. Such reverse mastery is accomplished, moreover, in a newly clarified relation of unity with nature; man's unity with it becomes differentiated, interactive, experimental, and reflective.

Out of this situation of estrangement toward nature, an estrangement qualified by man's partial dominance over it and unity with it, attitudes and roles of mastery and slavery arise in society. Faced with continuous

threats of enslavement to the scarcity of nature (history shows that mankind has never been free of famine and disease, and today's world is no
exception), men both cooperate and compete in the desire to combat the
threat. But the competition is turned inward against the species; classes
arise to command and use the labor of others. Again, a dialectical twist
occurs: the cooperative laborers whose labor-power is owned by the ruling
classes appear to be in subjection, in slavery, to both social and natural
powers, for they have the least goods from nature and the least control
in society. In hidden fact, however, they are, so far as their techology
extends, the masters of nature, and therefore they implicitly hold the
power of life and death over their putative social masters. At this stage
of deep contradiction, it becomes clear to anyone conscious of the contradiction, of this unacknowledged mastery of men over both nature and
society, that the next stage in human history, if there is to be a stage at
all, must inevitably be that in which this social mastery is brought to
consciousness and realized in social life. The laborers must become class-
conscious and in the class struggle must succeed to power over the capitalist ruling class.

The slave may undeceive himself. He may discover in the act of self-
consciousness that he is not really the determinate, enslaved, sensuous
being that he appears to be but in fact possesses a potential freedom as
evidenced in his labor with and over nature. That is, in his relation to
nature man first lives in an unconscious unity; individual desire has not
yet differentiated him from it. Then he passes into a period of historical
scarcity in which he either settles for relative slavery to the dictates of
nature or else achieves a relative dominance over it as ruthless master.
In either case, he is in alienation from it. Deceived by appearances, he
believes that nature is his enemy, though as we have seen, his labor upon
it liberates him from its dominance to some degree, even if he is still
dominated by a social master, and his dominance over it involves him in
a new relation of dependence upon it. At last, when a sufficiently large
number of persons attain a collective self-consciousness and understand
through liberating practice the universality that binds them to one another
and to nature, when they achieve an economic and spiritual relation to
nature that requires neither slavery nor mastery toward it—then man
creates a new relation to nature. That relation both detaches him from
it and attaches him to it (as to other persons) in new ways not hitherto
experienced in human history. It transcends the old relations of mystical

union, religious propitiation, and class exploitation because the social and natural conditions for such relations have been transcended.

For Marx and Engels the liberation of humanity, i.e., mostly laborers, from alienation within both society and nature must come through the mighty historical act of the class struggle. This struggle, which requires the unity of mass consciousness and political activity of laborers, is itself an ecological movement, for it consists of individual persons who enter into new relations of cooperation and mutual aid with one another in their struggle, and who strive toward new relations of cooperation and mutual aid with nonhuman nature. Such relations are "new" in that they pass beyond the old historical relations of the ruthless competition of each person with all society and of the cold-hearted, short-sighted, heedless use of nonhuman nature. The creative emergence of the proletariat in history is a movement of a new species toward a new society in new relations with its natural environment. This is a qualitatively new humanism in history, a proletarian humanism, and a qualitatively new naturalism, a humanistic naturalism. In it laborers, in thought and feeling and action, embrace their fellow human beings and the whole domain of nonhuman nature as their own. They protect and cherish the whole human-natural environment, providing conditions necessary to bring the whole ecological system of the planet to its creative fruition. The inevitable pathway to this new ecological system is the transformation of capitalism into socialism under the leadership of a proletarian political movement.

V. Transition from a Capitalist to a Socialist Ecology

How do the theory and practice of Marx and Engels toward nature differ from those of the capitalist? What are the breaking points, the places at which their theory and practice must stand in stark and active opposition to capitalism and thence move on toward a new ecology? Marx and Engels agreed with the capitalist "stratagem" to "subdue" nature for "human requirements." But they disagreed by claiming that (1) the mastery of nature should benefit all people and not just a small ruling class, (2) the mastery should maintain the dialectical balance of natural ecology in harmony with human needs instead of destroying "our one and all" by making the earth "an object of huckstering" (see

Selection VIII B), and (3) the mastery should be qualified by a theoretical understanding and esthetic appreciation of nature which are the opposite of capitalist "contempt." In their efforts to construct a science of political economy and in their daily political struggles for socialism, the ecological and esthetic dimensions receded into the background, though as essential dimensions of man they were assumed from the beginning as necessary parts of the final goal. Since 1917, moreover, socialist states, starting from a poor and undeveloped economic base, have been continuously subverted and encircled, and repeatedly attacked, by capitalist states. The emphasis on action and mastery of natural and social problems bearing on food, raw materials, industry, defense, and the like has reflected these conditions of economy, subversion, encirclement, and attack. Even so, the love and enjoyment of nature appear abundantly in artistic expressions in socialist countries, while ecological concerns and the esthetic mode of perceiving nature have not until the last decade or so figured prominently in the philosophical literature.

The usual capitalist is a short-sighted, profit-hungry predator on both man and nature, a subjectivist in his value-theory, an objectivist only insofar as his knowledge of the objective world enables him to satisfy his subjective desires. For him, both nature and other men are there to be negated. In contrast to this attitude of unrestrained mastery over nature, Engels wrote:

> . . . we are reminded that we by no means rule over nature
> like a conqueror over a foreign people, like someone stand-
> ing outside nature—but that we, with flesh, blood, and
> brain, belong to nature, and exist in its midst, and that all
> our mastery of it consists in the fact that we have the advan-
> tage over all other creatures of being able to know and cor-
> rectly apply its laws. (See Selection VIII F.)

The more we know and control the natural consequences of our productive activities, he argued, "the more will men once more not only feel, but also know, themselves to be one with nature." (See Selection VIII F.) The word "feel" is important here, as we have observed earlier, since Marxists have sometimes seemed to take an exclusively mechanistic, quantitative attitude toward nature—overlooking or denying its intrinsic qualities or else employing the traditional materialistic dualism of primary and secondary qualities.

Here, as elsewhere, Engels reproves the capitalist practice of acting as "conqueror" over nature but retains the image of "master" based on rational understanding of nature and rational practice toward it. ("Rational" here as always for Engels and Marx means dialectical understanding of man and nature for the purpose of promoting the welfare of mankind in the long run.) Freedom is man's insight into the necessities of nature's laws and his consequent practical adjustment to them for the sake of satisfying human needs. The word "mastery," however, must be understood in context. Engels emphasizes that we are immanent in nature. We do not stand outside and over it, as a man stands over an automobile to repair it; we "belong" to it, in the sense that we have an organic, functional, and qualitative connection to it. For Marx and Engels, man is profoundly dependent on an antecedent nature. Interacting with it to survive and live well, man becomes independent through reason and rational practice, but the independence is always conditioned by its sources, surroundings, and goals within nature.

From the viewpoint of an ecologist who grasps *sub specie spinozae* the complex of natural relations, the term "mastery" may seem to overstate the case for man as an independent variable, as it may for a socialist society that has achieved a felicitous degree of harmony with nature. But from the viewpoint of a hungry man in a society of cruel competition and scarcity, the problem of "mastery" over the environment, by hook or crook, will normally appear as the paramount one. Having in mind man's needs and the possibilities inherent in industrial technology for satisfying those needs, Marx and Engels retained the term. They rejected the machine-wrecking Luddites, the agrarian utopianism of men like Fourier, and the indifference, skepticism, and disdain of men of "culture" toward the democratic promise of industrialism. Since the Renaissance, the term "mastery" has expressed a revived faith of man in man. Marx and Engels, moreover, shared the faith of a spectacularly successful nineteenth-century capitalism in material and technological progress—though in contrast to capitalism they called for a *social mastery* of nature for the sake of *man*. They did not and could not foresee the rude jolts that the "ecological crisis" has given to this faith, especially as held by those who had hitched their wagons to the declining star of capitalism. Had they done so, we may reasonably suppose that they would have declared themselves more vigorously and explicitly on the ecological side of their man-nature dialectics and on the healthy social and natural order that must be the only workable alternative to a planet plundered by capitalism for 1000 years. Even so, as we have shown, they set down in their time many passages on such man-nature dialectics, leaving no doubt as to what their ecological position was then and would be today.

The term "mastery" in the expression "mastery of nature" is ambiguous. It means (1) a high degree of skill or knowledge; (2) intellectual superiority or some degree of practical power; and (3) nearly complete or complete determination or domination, as that of a master over a slave. Few men would deny that some men possess mastery over nature in senses 1 and 2. As for 3, it is doubtful that men's domination of themselves and their environment can be even "nearly complete" when that environment, with its physical, biological, and social laws, its massiveness, complexity, and historical growth of structures (in atom, gene, human body, social institution), determines the existence of all men from their conceptions to their deaths. The illusion of such "mastery" comes from ignoring the pervasive working of such laws and the evolution of man as a living material form over a period of 3 billion years, delicately and complexly adapted to the evolving environment of nature and a living community. For example, persons from capitalist as well as socialist nations speak of "the conquest of space." It would be more accurate to speak, dialectically, of man's mastery-through-submission, of the Hegel-Engels' freedom that is "insight into necessity," or of the Baconian dictum that "nature to be commanded must be obeyed."[157] Some so-called Marxist formulations of "mastery" of nature are too frequently infected with the capitalist attitude and practice of possessing, grasping, and eventually destroying. They are not sufficiently humanistically differentiated from the "greed" that Marx and Engels so often denounced. The danger is an unguarded (undialectical) emphasis on man's differences from and antagonism toward nonhuman nature and toward his enemies in human society. The danger is blind ruthlessness toward both nature and men. On the question of mastery, Marxists must continue to make clear, as Marx and Engels did, that their ecological position is the very antithesis of that of capitalism: governed by care and not avarice toward man and nature, generous and not possessive, planful for nature and society.

This error in viewing man's relation to nature sometimes springs from Marx's and Engels' own emphatic and unguarded statements about man's "appropriation" and "mastery" of nature and sometimes from a misinterpretation of their formulations. An example of the latter is Engels' *"die Einsicht in die Notwendigkeit,"* an expression which he borrowed from Hegel and which is variously translated as "the appreciation of necessity" (see Selection IV H) and "the understanding of necessity." Such translations suggest an attitude of passive acceptance on the part of man toward an unchangeable reality—a contemplative, stoical man mechanically geared into the machinery of nature and impotent to alter its inevitable course. This is a positivist picture which sharply splits off the world of "facts" from the "emotive" evaluations, purposes, and actions of man.

But such a picture is the antithesis of Engels' meaning. *"Einsicht"* means insight[158] which fuses both an intelligent, sensible understanding of things and a good judgment or sensible evaluation of them. It means knowledge of the elements and relations of the objective situation as well as a selective and integrative act of evaluation which includes judgment on the relevance of such facts to human purposes. It means wholistic cognition bearing on the fulfillment of human needs, not atomistic contemplation that leaves needs and action out of account. Otherwise, freedom with its "insight" could mean only an abstracted or haphazard accumulation of data which not even dilettantes engage in. Situations, moreover, are not impermeable. They yield to the intervention of human action— but only on the condition that the action is informed and guided by the character and behavior of things. Thus, "insight into necessity" has a dialectical meaning that indicates the interplay of perceptually based, conceptually informed judgment on the objective situation. Only as thus conceived can man's knowledge be ecological or have any relevance to ecological problems. Indeed, only if man is from the start an ecological creature, dependent upon external nature and influencing it, can dialectical knowledge arise. For such knowledge is constituted by the dialectical transactions of the organism with the environment as these transactions are brought to consciousness in the form of symbolic patterns signifying the regular behavior of things and their relevance to human needs, actions, and values.

The term "values" indicates the human effects of actions and events on fulfillment or deprivation. (When "facts" impinge on human needs, they take on, in that relation, a value-character, positive or negative.) Scientists and politicians speak glibly of avoiding the "side-effects" of nuclear explosions or of drugs like thalidomide. But *calling* effects like leukemia and bodily deformity merely "side-effects" does not alter the lethal and monstrous character of the effects on human beings or the long-range effects on the economy of nature. (Of course, some scientists deliberately aim at the promotion of human terror, suffering, and death, like those U.S. scientists who for the Vietnamese people devised the razor-sharp splinters of the "lazy dog," the "pineapple bomb" of maiming pellets, and many others.)[159] Science is not a collection of isolated propositions, each contemplated like an autonomous stone, each applied to its appropriate piece of the world, in invincible ignorance of the environing universe and its truths. Instead, it is an uncompleted, growing tissue of interconnected truths, reflecting with approximate fidelity the interconnection of facts in space-time.

Real, living science is in fact much more than such an abstract scheme of interconnected truths. It is the activities of a large number of men and women, a community of investigators in the space-time of our planet, who are acting on the world of things and processes in experimental, habitual, and cooperative ways so as to create technologies that improve or impair the community of man. The community of man, moreover, is a community of interconnected, interdependent individuals living in interaction with other parts of the biosphere of the planet. The effects of actions of individuals and of the species are far-reaching. The scientist seeks to relate his own knowledge and activity in an integrative way to the needs of the members of the species and to the laws of the biological and physical environment, to satisfy both the demands of human beings and the necessary conditions confronting them in their planetary environment. An isolated scientist who takes no heed of the consequences of his acts, either for the sake of "his own truth" (a fiction) or for the sake of humanity, or a scientist who isolates his truths from other truths, or a scientist who isolates his activities from the activities of other scientists or from the needs and fulfillment of the members of mankind—such a scientist is a contradiction and a perversion. For if he is committed to a single truth, a scientist must in principle be committed to all truths, since all truths form an interconnected system. Since the meaning and value of any truth and all truths must be the meaning and value of truth *for man* and man's realization as man—man not only as an individual but as a species of interconnected members in a nature of interconnected things—then to seek, possess, or enjoy any truth in isolation from the needs of human beings as individuals and as species is a criminal dismembering of science from man and nature, a flagrant betrayal of science, man, and nature. The enterprise of theoretical-technological science is *par excellence* a *human* and *ecological* enterprise. A reverence for truth that is true to its task is also a reverence for man and nature.

As among the mechanics of the inventive nineteenth century, so among the twentieth-century engineer-politician-military complex, the mechanical Americans who are so hung up on athletics, the flag, spaceships, and computerized warfare—i.e., the ethos of "winning"—there is the widely shared metaphysics of triumphant manipulation, of the superiority of man in the competition of the cosmos. The theoretical adjuncts to this technocratic enterprise are the "think tanks," their geniuses solemnly engaged in their gamesmanship, their strategies of offense and defense, their calculations of first-strike capabilities and megatons and megadeaths, their space fantasies and space Armageddons. It is a game of power, of power over the competitor, and ultimate power means ultimate destruction—of man, of nature. Has not man, with his atomic physics, his nuclear reactors, his Saturn rockets, and his flights to the moon and the planets already outwitted nature? After all, who needs that weak, inefficient, organic system called Mother Nature, when politicians can finance and engineers can contrive a self-contained, inorganic space capsule for the astronaut who, with his own "life-support system," can then defy gravity, history, and civilization for a transcendent landing on the moon?[160] Was not that, as President Nixon said, "the greatest week in the history of the world since creation"?

This is the same attitude of technological self-centered arrogance that not only reared the most highly developed industrial society in the capitalist world but also in the space of two centuries introduced extensive deforestation, deep plowing, poor land management, the destruction of 280 million acres of crop and range land, the irreparable erosion of 100 million acres of land,[161] the extinction and near-extinction of many animal species, and the socially planless exploitation of minerals. In the past a certain amount of technological chutzpah could pass without having immediate and obvious effects on man. Men might overcultivate the soil until it was infertile, overgraze the grassland, overload the rivers and harbors with waste, and overpopulate the cities. But there was always a chance for some of the rich or enterprising to shove off or hit the trail toward some beckoning frontier, toward a new ecosystem where a fresh start with fresh resources might be made. Now, however, the domain of fresh resources is rapidly diminishing and in some places is threatened with extinction. Where technology is not adequately advanced or organized for social benefit, capitalist interests sweep aside ecological considerations in favor of quick profits. In the United States every year nearly 1 million acres of land,[162] many of them green, give way to the black death of asphalt and concrete and other urban accompaniments, the epidemic plague carried by packs of "developers." After extensive pollution, life is only slowly reviving in Lake Erie. The waters of Lake Superior are not fit to drink, and the waters of southern Lake Michigan are polluted daily by industrialists who find it cheaper to pay the fines rather than install equipment to prevent pollution at the source. "Imminent biological death" haunts the lower Mississippi Valley and "numerous lakes, streams, and coastal waters."[163] The tropical rain forest of the Amazon River basin, the Congo River basin, and the Malay Archipelago, "the most complex ecosystem on the earth," abundant and diversified in plant and animal species and their relations, is being "developed" by lumbering, farming, and plantation interests and may virtually disappear by the year 2000.[164] The oceans, teeming with marine life and supplying 70 percent of the planet's oxygen, are exploited with little overall plan (will the whales survive?), are increasingly polluted, and are endangered by a growing marine industry. The exploration of space has been a mixed bag of national competition, propaganda feats, scientific study, and international cooperation. Indochina became for a time a lucrative frontier for American imperialism and its genocide and ecocide.[165]

For some decades now, scientists have prognosticated in various ways the "energy crisis." Marx's and Engels' description of capitalism was an account of how the production and distribution of energy are frustrated by a system of private property relations and exchange, so for them such a crisis would be credible and predictable. Industrial capitalism and monopoly capitalism have been built upon a foundation of coal: in the past 110 years, 19 times as much coal was produced and consumed as in the preceding seven centuries. But the production and consumption of coal and other fossil fuels are increasing exponentially, and 80 percent of the total finite supply can be expected to be produced in the period 2000-2300.[166] Such limits have been known for some time. But what does capitalism do? The monopolists of fossil fuels continue with "business as usual," and the governments with their fossil mentality continue to serve as a "committee" for managing the affairs of this ruling class. (The oil monopolists contributed almost $5 million to President Nixon's 1972 campaign and were repaid by superprofits and government noninterference.) The production and distribution of energy from fossil fuels and other sources (solar, geothermal, tidal) are thus determined not by long-range consideration for human needs and ecological criteria but by "costs," i.e., surplus-value that accrues to the private entrepreneur. Hence for the next fifty years U.S. society will use dirty fuels (coal and petroleum); only massive democratic demands from the people can move it in the direction of humanistic planning of energy resources.

Specific perils of an illusory notion of man's "mastery of nature" have emerged in the last thirty years. The technology of atomic fission has given mankind the power of absolute death over all living matter on the planet, as well as the power to inflict genetic and somatic injury from atomic-bomb testing by national governments (still going on) and nuclear accidents (increasing in probability). Other new technologies—control of genetic materials, organ transplants, devastating methods of warfare, automation and cybernation, contraception and abortion, drugs for altering states of the body and consciousness, prolongation of life, monitoring and surveillance, concentration and control of information, mind-manipulating media—all have put into question the values of such extensive power. The technologies of modern industrial societies introduce new elements into the cycles of nature and accelerate them, with harmful consequences to man. Specifically, inorganic materials such as metals, nuclear waste, and concrete, used and produced in industrial process, get into natural cycles and cause injury and death to living organisms;[167] man's release of stored energy in fossil and nuclear fuels accelerates the cycles of nature, and the byproducts (heat, carbon dioxide, water vapor, carbon monoxide, sulfur oxide, hydrocarbons) interact with the cycles;[168] and man's technological agriculture produces soil erosion, dust pollution, water-logging of roots and salinity, pollution of drinking water, destruction of marine life, and injury and destruction of insect and plant life.[169] Harmful hydrocarbons from the exhaust of jet airplanes, insecticides like DDT, herbicides, detergents, radioactive substances, and other materials, which are more profitable for private enterprise to produce than their harmless alternatives, have gotten into food chains and have adversely affected living things and the web of life in which they are sustained. Of some 47,000 synthetic organic chemicals for use in industry (presumably in the United States), only 8,000 have been tested even superficially for their consequences on man and nature.[170] In the United States, 500,000 to 600,000 chemicals are produced (increasing at an annual rate of 500) and their effects on man's health are "largely unexplored."[171] Besides the toxic exposures of chemicals in factories and mines[172]—as well as pollutants in the atmosphere, water, and soil—food additives, new drugs, and other chemicals in the environment possess the potential for injuring or destroying human life. Although we do not yet have "hell to pay," we are burdened with a mounting debt to nature (to continue the bookkeeping figure), which we are blindly bequeathing to unborn generations. Purely commercial interest or an engineer-politician-military mentality such as described can have no appreciation of this debt because it has no sense of the interconnections of man and nature. "Behavioral engineering" is on the scene and biological engineering has appeared on the horizon.

While Marx and Engels did not of course anticipate these specific perils and problems of late twentieth-century industrial capitalism or the problems of twentieth-century socialist societies, they did set forth an explanatory and predictive scheme of the structure and dynamics of capitalism by which the general causes of the perils and problems might be understood and avoided. We have already expounded this scheme in detail, so we will not repeat it. The principal concept in the scheme is that an economy based on capital simultaneously organizes and releases great productive power but drives in the direction of profit-maximization. These two tendencies, respectively centered in the working class and the capitalist class, prove to be antagonistic and incompatible. The drive for profit-maximization inherently rejects the primacy of safeguarding human life and meeting human needs, overall social planning, and long-range, ecological consideration of the social and natural consequences of decisions in the productive process. Hence the productive power increasingly runs amok, issuing in disorder, insult, injury, and destruction in both the human and the natural environment. The way to avoid these bad consequences is to change the system to one that is purged of destructive class conflict and the private appropriation of social wealth, one that is humanistic, planful, and long-range in its social and ecological orientation. Where such a change has been made, in socialist economies, the dialectical relation of man to man and of man to nature, in the light of our new scientific knowledge and our new problems and opportunities, must be taken with new seriousness.

To understand where we are in the present economy and ecology of the planet, what our major problems and opportunities are, what the role of Marxism is with respect to those problems, and what the future may be and ought to be, it is necessary to take a brief look at the evolution of the system of living things on the planet, including the evolution of human society.

We are now at the climactic and critical stage of a very long ecological process. Some 4.5 billion years ago our planet was formed out of gaseous matter, and 3 billion years ago the first heterotrophic bacteria emerged, followed by autotrophic, photosynthetic organisms. These organisms marked the triumph of antientropic forces, the evolution of an organization of matter which transcends the physical drift, collision, and waste of matter and lifts it up into a system of mutually reinforcing and mutually enhancing parts, a system that, against the blind movement and dissolution of energy, creates, preserves, increases, and perpetuates energy. For 3 billion years this system of living matter (for it was and is a single ecological system) has evolved, and the fitness of the environment for it has evolved with it.[173]

This system has two important characteristics from the point of view of objective understanding and man's practice. One is its selective, definite, exclusive structure. The material universe as now known consists almost entirely of gas: 99 percent of it is composed of hydrogen and helium. These and other gaseous materials are organized into thermonuclear reaction systems in the form of stars, whose galactic clusters send out their incalculable energies, radiant and nonradiant, into the vast intergalactic spaces. Subnuclear transformations of matter in the depths of quasars and exploding galaxies release still more powerful energies into space. Yet, in the midst of such pervasive violence and chaos, a stable order of material atoms exists in limited, isolated portions of this thinly settled universe. For there are more than 100 known atomic elements, themselves an array of selected, synthesized realizations from among the possible combinations of more than 50 known subatomic particles. Such elements conglomerate and unite to form the stable solid matter of which solar planets like our own are made. The crust of our own planet is composed for the most part (98 percent) of only eight of these elements—oxygen, silicon, aluminum, iron, calcium, sodium, potassium, and magnesium. The sea water covering most of the earth's surface is almost entirely composed of the two elements hydrogen and oxygen—a very simple medium in which and out of which our ancestors took their rise and which is still the medium in which the solid cells of our bodies are bathed.

Living matter is now known to require a complex arrangement of 24 of the more than 100 elements, with hydrogen, oxygen, carbon, and nitrogen making up more than 87 percent. Protoplasm is thus not a replica of the physical universe. It creates itself by utilizing only a small proportion of the building blocks available, and it selects and organizes only those kinds of atoms and molecules relevant to its structure and functions and shapable into living tissue. Thus, water is a solvent base to which certain compounds can be usefully related; carbon is stable and capable of forming long chains and hence organic compounds; and the most abundant atoms in living matter—hydrogen, carbon, oxygen, and nitrogen— are small, light, and able to share their electrons with other atoms and thus lay the base for stable molecules. It is the cooperative elements and not the "noble" individualistic ones that make life possible.

All this means that living forms during millions of years of evolution have created very specific systems of integrity and of exchange with their environments. These systems are imperiled or destroyed when the materials that they take in are excessive or deficient for the needs of the systems or are incompatible with their integrity. Such, for example, are mercury and lead, two common byproducts discharged into the environment by industrial activity—though certain exceptional organisms, like the tuna, swordfish, and percomorph of the sea, which concentrate high contents of mercury, appear to do so as a long-time adjustive response to an environment naturally rich in mercury. Strontium-90, a rare substance in nature but a product of man-made thermonuclear fallout, is easily mistaken for calcium in the marrow of man's bones and causes disease and death. Furthermore, there are fine adjustments of viruses and bacteria with other organisms;[174] insects are specially adapted to particular plants and animals and maintain the equilibrium among living things. There are complicated food chains the disturbance of which has ramified consequences. To ignore the intricate machinery of this long-evolved and well-adapted living system, and to impose upon it blindly the feats of man's engineering "wisdom" are folly.

The other characteristic of the living system is irreversibility in time.[175] With the rise of organisms, a certain directiveness appears in matter.[176] Directiveness means selection with regard to both ends and means, selection from among the material possibilities of realization inside and outside the organism. But matter is in constant change, and the process of evolution in the individual organism, the species, the system of living beings, and the whole planetary system is a process of selection from among present material possibilities. It is the building up of a temporal structure of levels from the simple to the complex, in which the later levels depend on the earlier ones. Individuals and species normally either evolve and stabilize, or die; they do not degrow or self-destruct and then regenerate. Once a species on this planet is gone, it is gone forever, because the unique complex of genes and environmental conditions (with their space-time of millions of years) that produced it are gone forever. Once an ecosystem (such as the Scottish Islands,[177] the Tennessee Valley,[178] or one of many other ecosystems on the planet) has been seriously disturbed, it may be partially restored or replaced with another ecosystem, but it cannot be returned to its original form. Urban consumers oriented to immediate satisfactions and accustomed to the convenient replacement of any consumed object, such as food or a television set, are apt to take a cavalier attitude toward nature in this regard. They may imagine mountains to be mere piles of dirt and rocks, as easy to make or remove as the ski areas that displace them.

Plants and animals have evolved on the planet in the teeth of hostile forces of inanimate matter—the heat and cold, the fire and frost and flood, the cataclysms of earth and sea, the inertia and chance drift of inorganic things. The emergence of the DNA and RNA molecules, with their power to direct growth and reproduce, made possible a precarious conquest of this inertia and drift. The photosynthetic power to transform water, carbon dioxide, and the radiant energy of the sun developed in plants. The power to metabolize plant protoplasm into animal protein developed in animals. These two forms of life thus overcame the dispersion and wastage of inanimate matter. Negentropic,[179] they gathered and organized energies into growing systems of equilibrium—in organisms, species, and interspecies relations—and gradually created a new system of nature, a nature increasingly dominated by organisms interpenetrating and transforming their environment.

Single-celled organisms learned that if they cooperated they might perform their necessary functions more efficiently and survive more successfully. The consequence was, first, loose aggregations of unicellular organisms, such as ciliate infusoria, which conditioned their common environment in certain favorable chemical ways. Then came colonial protozoans like sponges and slime molds, demonstrating the cooperative organization of single-celled organisms from erstwhile separate individuals. Finally, most multicellular organisms evolved from unicellular organisms that specialized in their replication and developed integrated systems with elaborate division of labor among specialized groups of cells for sensitivity, locomotion, digestion, and the like. The opportunism of evolution pro-

duced a great variety of such structures with varying degrees of power in
the production and organization of energy. (For example, planktonic
algae convert solar energy into chemically bound energy with 10 to 20
percent efficiency, whereas corn is only about 2 percent efficient.)[180]
In evolutionary history, we can discern a number of stages along this line
of advance, but for our purposes here we note the mammalian sociality
and the primate family.[181] Under changing climatic conditions these
forms of cooperation proved much more efficient in the use of energy
than the reptiles' mode of life which emphasized reproduction in large
numbers, quick early growth, and individual survival.

The ancestors of man were advanced primates whose rapidly developing brain, hand, upright posture, and limbs became the biological basis
for tools, language, and effective social and family life. With hunting tools
and social speech, energy in the form of food, shelter, and clothing could
be obtained with relative ease. (Bushmen today spend only one-third of
their daytime getting food.) Thus hunter-gatherer hominids began slowly
to transform their physical environment. Probably as long ago as 25 million years, they acquired the use of primitive tools as digging sticks and
clubs.[182] For millions of years they survived, wandering in small groups
over the savannas of Africa in search of small game. We know now that
stone tools were being manufactured in northern Kenya more than 2.5
million years ago by *Australopithecus*,[183] though they may have been
made and used long before that. The mastery of fire as a tool and weapon
appeared 0.75 to 1 million years ago,[184] and with fire man modified the
quantity and movement of game, cleared brush and forest, cooked, and
congregated at night. The use of fire is correlated with the rise of the
various species of *Homo.* From *Australopithecus* to *Homo Erectus,*
a period of 2 million years, the brain size dramatically doubled. Such a
brain was the concomitant of cooperative big-game hunting (requiring
eye-hand-speech coordination) and the result of the new meat diet, and
was associated with the loss of estrus, the close bond of male and female,
infantile dependency, and intimate family life.

With the recession of the last glaciers more than 10,000 years ago, and
with the improvement of his tools, man commenced to move toward a
new stage in his evolution. Aided by the warming of temperatures, he
was able to exterminate many birds and mammals, such as the mammoths
and mastodons. Through the controlled use of fire, he emptied the bush
and forests of game and rendered river valleys ready for the primitive

plow. By the time of Christ, half of the hunter-gatherers had been displaced by an economy based on domestication of wild nature, i.e., plants
and animals. Tillers and herders dominated the landscape. They had laid
the literal ground for the food-producing revolution that would usher in
the most important change in genus *Homo* since it emerged from the trees
more than 25 million years ago. The creativity of social labor with its
surplus production made possible the support of a population not directly
engaged in the production of food. Specialization of labor arose, and
with it classes and the first empires.

During the brief period between 5000 and 3000 B.C., momentous
technological changes occurred: the invention and widespread use of
artificial irrigation, the plow, the harnessing of animal power, the sail-
boat, wheeled vehicles, orchard husbandry, the metallurgy of copper,
bricks, the arch, glazing, and the seal.[185] Such changes paved the way
for the urban revolution with man's first civilizations, surplus food to
sustain large cities, a consequent spurt in population growth, crises of
overproduction and surplus population, and eventual collapse of the
civilizations because of their inability to resolve the contradiction between
their productive means and the demands of the population that they had
generated. (Wars and colonies in quest of food and slave labor power
were ineffective expedients for solving the problems.) Nevertheless, human
technology had accomplished on a large scale what the first cells had
accomplished on a small scale: the negentropic production, conservation,
and transmission of surplus energy. But society could not organize itself
to manage and utilize the new energy in socially useful ways, and so the
crises of these early civilizations were followed by relapses into simpler
societies.

A second wave of the technological and urban revolution commenced
a thousand years ago in the West. Its technological improvements arose
from a base of common technological practices of medieval society, drew
its intellectual stimulus from sources in the previous revolution made
available by translations from Arab and Greek scientific writings, and
derived its motive power from new machinery for extracting and using
energy. The movement of this wave can be plotted by its successive
energetic crests, by the increasingly shortened periods required for a
doubling of the world's population: 1 A.D.-1650, 1650-1850, 1850-1930,
1930-1975.[186] Such periods correspond roughly to stages of energy-use:
wood and water were the principal sources of power from 1000 to 1750;

iron and coal came into wide use in about 1750; and electricity and alloys were developed and applied in 1830.[187] The last stage, the one we are in now, is marked by the systematic application of chemistry and atomic physics to industry and by the development of automation.[188] We live in an era of agricultural chemicals, pesticides, massive pharmaceutical industries, synthetic rubber, plastics, instrumentation, automated machines, and factories, an era in which nuclear power is supplementing power from fossil fuel and hydroelectric sources.

Marx and Engels recognized that the system of capitalism was the carrier of this technological revolution but at the same time the impediment to its worldwide unfolding for unrestricted human benefit. By its production of new forms of energy, capitalism has brought forth the promise of life and creative leisure on a species-wide scale. But by its forms of economic and social organization, its private ownership, it has prevented the fulfillment of this promise. The egg of self-defeat laid in its blossom comes from the moth of its own making: it applies universal laws of energy production for private ends. It assumes that science, whether the science of society (political economy) or the science of nature, can be utilized for the benefit of a particular class to the harm of the working class and the vast majority of the human species as well as other species, without serious consequence to society or nature. (That is, it defines "humanity" in a very narrow way, and it ignores "nature" as a concept or reduces it to what Whitehead has called "vacuous actuality.")[189] But every act of capital has proved these notions to be false. Science by definition is planetary and universal; the ultimate laws that it describes obtain for all men and all species and all natural conditions under specified conditions on the planet. Hence, those who apply general and far-reaching scientific truths in the control of nature and social conditions produce altered conditions that affect not only themselves and their class but also all men of the human species, other species, and inorganic things. An alteration that is designed to limit beneficial effects to a few or to ignore harmful effects for all is therefore destined for failure. Capital has sown the wind of profits for itself, and has reaped the whirlwind of miseries for mankind and nature.

Physics has long taught the law of the conservation of energy, and the ecological law that "what goes into a system must eventually come out"[190] is simply a translation of this law into systemic terms. The law applies to both natural and social systems. It looks simple enough, but the fact is

that in the history of capitalism it has been repeatedly regarded with contempt. For example, early nineteenth-century capitalism, based on extractive industries without replenishment, took no care to redistribute human
waste back to the soil or to restore forests. The materials extracted from
the earth by man today—iron, phosphorus, lead—are washed into the
oceans in the amount of millions of tons each year, without recovery.[191]
Only recently has capitalism shown any interest in the less profitable conservation and recycling of such materials. Capitalism has always sought
new land frontiers from which to extract materials, and now that the
land it has used and abused is becoming exhausted it is intensifying its
competitive exploitation of the seas and the air. (Who will regulate the
commercial extraction of oxygen from the atmosphere? What will be a
"fair profit" for that? Will entrepreneurs open "service stations" to sell
oxygen to the suffocating?) Simultaneously, capitalism has extracted
more and more energy from the laborer, but has left him unrecompensed
in ways that do not permit him to use the values that he has produced.
In its advanced stage of commodity production, capitalism exploits the
consumer, probing into the strata of his desires as a mining company
probes into the deeps of the earth. First, there was motivational research;
then, subliminal research; and now, psychographics, which describes and
exploits the moods of the buyer. However, extraction of energy from
both nature and personality has not reckoned with the consequences.
A social system, like a natural one, can become exhausted, polluted, and
fragmented. From so much depletion, it can reach a point of rebellion in
society or unresponsiveness in both society and nature.

Marx studied the work of the agricultural chemists Justus von Liebig
and Christian Frederick Schönbein,[192] and it was Liebig especially who
emphasized the importance of returning organic matter to the soil. In
1868 Marx wrote Engels of his reading of the work by K. N. Fraas,
Klima und Pflanzenwelt in der Zeit, eine Geschichte beider (1847).[193]
Fraas contended that through cultivation and then deforestation man
had altered the vegetation pattern, developed in response to climatic conditions, and in consequence the climate too. Marx noted that the devastation thus wrought by the intervention of man in nature was in fact a
bourgeois enterprise and that Fraas's observation of the results of cultivation not consciously controlled is "another unconscious socialist tendency."
Marx and Engels were confident that, given the intelligence and technological
power of man, the untapped resources of nature, and the creativity of

socialist organization, even the enormous wastage of capitalism could be overcome. This optimism was shared by many in the twentieth century, including the eminent Russian scientist Vladimir Vernadsky. Today, Soviet scientists hold a similar view, though it does not gloss over the hard facts of environmental degradation and calls for international cooperation.[194] For example, the hard fact of the declining volume of nonrenewable forms of natural wealth, they argue, does not mean, as some U.S. ecologists contend, a decline in the capacity to satisfy various human needs. For under socialism that capacity *increases,* through a rise in the effective use of each resource, through the use of new resources, and through new approaches to satisfying needs.[195] Thus, with socialist organization, the *social* wealth of reason and cooperation greatly increases in ways that counteract the decrease of natural wealth that occurs with the depletion of metals, for example, and with the production of nonbiodegradable substances such as synthetic fibers, detergents, and fossil and nuclear fuel.[196]

However, Marx and Engels could not predict the extent to which modern technology under the dominance of capitalism would corrupt the systems of nature and society. They cited the depletion of the agricultural soil and natural resources, and in dealing with the problem of population and resources, so critical today, Engels believed that "too little is produced"[197] and did not prescribe what to do in the event of a shortage of resources. Similarly, Marx and Engels rejected Ricardo's theory that the fertility of the soil decreases as population increases. As early as 1843, Engels had remarked on the role of science in the progress of agriculture, a science whose increase is "at least as much as population"—"and what is impossible to science?"[198]

It is important to place this optimism in its context. Between 1750 and 1850, the population of Europe almost doubled as a consequence of industrialism and the reduced impact of the traditional killers, war, famine, and disease. At the same time the new food supplies (potatoes and maize) were insufficient for the new millions, and even widespread celibacy and infanticide failed to check the growing numbers. Marx and Engels were witness to this population explosion and the miseries attendant upon it, and they undertook to answer its gloomy prophets like Malthus. Their chief answer was that industrial production, scientifically and socially organized, could sustain an expanding population. They foresaw a relatively early transition to such an economy and did not anticipate the complex of environmental problems that an exponential growth of capitalist produc tion would bring in the next century or the limits that finite resources would pose to such growth. But they were correct in rejecting Malthus' and Ricardo's theory of decreasing fertility of soil, for advances in agricultural science and machinery have greatly increased the fertility of the soil and have made possible the increase in population.

Today environmental stresses derive from a large number of substances introduced into the system by advanced technology: pesticides, heavy metals such as mercury, carbon dioxide, sulphur dioxide, suspended particulates in air, oil spills, waterborne industrial wastes, solid wastes, chemical fertilizers, organic sewage, nitrogen oxides, storable radioactive wastes, tritium and krypton-85, photochemical oxidants (smog), hydrocarbons in air, carbon monoxide, thermal pollution, and community noise.[199] Though Marx and Engels could not know of these specific developments, they were aware of the general thesis, taken from Fraas and others, that changes such as climatic ones are the results of human activities on the land and forests[200] and that man has great power to alter and disrupt natural patterns.

Marx showed that surplus-value is the foundation of capitalism and exposed the absurdity (contradictoriness) of the belief that this key element in the system could be extended without disintegrative consequences. A certain quantum of labor goes into the system; a portion of it comes out as surplus-value and is reinvested in more productive machinery, which reduces the quantum of labor. Marx described the consequent disequilibrium—e.g., in the falling rate of profit—and the periodic breakdown of such a system in the form of hampered productivity and consumption. He argued for an alternative rational system that would bring the productive power into equilibrium with the consuming needs of the producers, and the system of social production and consumption into harmony with the resources and limits of nature.

The leaders of U.S. capitalism today, however, still operate on principles that try to defy ecological law. Like the European powers of the nineteenth century, U.S. capitalism must seek ways to rescue itself from the crises plaguing it every seven to eleven years. In a chain reaction differing from the economies of early capitalism only in its complexity, its lower cost technology (more efficient machines, more skillful workers) leads to increased output, more or less the same prices as previously, unsold goods, and crisis.[201] The circulation and consumption of the products of labor are occluded. But profits depend on labor, and ways must be found to revive production to meet the crisis. Marx gave an account of the ways in which such crises have been met by capital: destroying existing capital, waging overseas wars of conquest of markets, expanding capital investments overseas, establishing colonies. Lenin pointed out the role of finance,[202] and Keynesian public works have been used. Meanwhile, since its mercantile period capitalism has sought a net capital export in order to maintain and expand its profits. Such "favorable" balance of trade was not in fact a balance and, like all other policies, only perpetuated the unstable cycles of the system. The ideology that justified capitalism regards nature and society (as the Other) as either indifferent or inimical to individual man's purposes, to be seized, processed, sold at a profit, and discarded, without any accounting. That is why the young Engels could write: "To make earth an object of huckstering—the earth which is our one and all, the first condition of our existence—was the last step toward making oneself an object of huckstering." The contradiction is that if the Other—persons, nonhuman nature—is organically connected to us, then to exploit it is to exploit ourselves. Engels called this huckstering of nature "an immorality surpassed only by the immorality of self-alienation." (See Selection VIII B.)

Surplus-value is that portion of the value of a product which labor has added to it over and beyond the cost of production. It is a value contributed by the creativity of labor, a value grabbed, accumulated, and exploited by the capitalist, a value not paid for. That is why capital is said to "live" on labor as a kind of parasite. In a similar way, we can speak of the surplus value of virgin nature, which yields up its plants and animals, its waters and woods, its fruits and roots, its beauties and wonders, in ways that appear at first glance to require no price. Man simply finds such bounty, food and raw materials and forms of power, such as water and wind and fire, and utilizes them to his own benefit. Labor, of course, greatly adds to such natural value, but profit depends on it and ground-rent is a function of the unpaid productivity of the soil.[203] Capitalism treats nature with the same heedlessness as it treats men—except that, as political economy has shown, the capitalist considers it wise to pay for the subsistence of his worker until he can reproduce and raise up a replica and substitute, whereas most capitalists do not believe in the necessity of maintaining conditions requisite to the fecundity of a given sphere of nature. The capitalist *motif* of self-expansion, of mobility and conquest, and its contempt for conservative modes of production, have led it to reject the maintenance of

ecological patterns. It has simply torn down one forest, plowed up one plot
of land, and moved on to another. Nature gave and gave, freely and
abundantly; for the stored energy in the fossil fuels, the organic fertilizers
in the soil, the ores and woods and animals and vegetables and fruit trees
had been in the making for millennia, accumulating in superfluity. But in
time, despoiled to the point of destitution at the very sources of its creativity, nature gave less and less, and in some places it gave no more.
Surplus-value, unpaid for, became deficit-value. Capitalism had eaten itself
out of house and home, and had bitten and in some cases destroyed the
hand that fed it.

In such an economy ideology justified and guided practice. "The mode
of production in material life determines the general character of the social,
political and spiritual processes of life."[204] The Cartesian-Newtonian world
view of the seventeenth and eighteenth centuries sprang from and supported a society of expanding and prosperous commerce and banking.
The external world was seen as an extended world of things in space,
whose "primary" qualities of size, weight, and shape could be directly
observed, subjected to observation and mathematical measurement, and
utilized for business purposes. Optics and astronomy developed out of
the practical demands of navigation, and newly won leisure among the
wealthier classes made possible the construction of a philosophy of rational nature for which physics provided the chief inspiration.

The development of industrial manufacturing in the nineteenth century,
especially in England, stimulated the science of chemistry in the bleach-
ing and dyeing of textiles and, with the rapid increase and urban agglomera-
tion of population, gave rise to the beginnings of the social sciences. In
biology, opportunities for systematic research were hard to come by,
as the struggles of Pasteur and the independent means required by Darwin
illustrate. For most scientific work in the nineteenth century was devoted
to what had "immediate utility,"[205] that is, physics and chemistry, just as
mechanics had dominated the eighteenth century in response to the demands of the new manufacturers. Thus the thought of many intellectuals
in this period tended to remain under the dominance of the natural sciences, and when not religious these intellectuals tended to take an engi-
neer's view of the universe, even when, as in Herbert Spencer's case, the idea
of evolution was influential. The isolated and specialized character of
nineteenth-century scientific research also produced various asocial materialisms as well as countervailing idealisms that attempted to deal with the

value questions set aside by the sciences and these materialisms.

In the last thirty years, population increase and the problems of providing food and health care for large numbers have generated in the relatively well-off nations a corresponding increase in biological studies—biology of the soil, plant growth, animal production in agriculture, and commercial fisheries; medical theory and practice; studies of renewable resources; biology and industrial technology (in the industries of food production and storage, the pharmaceutical industry, fermentation and related industries, instrumentation, and pesticides); and environmental health.[206] All these sciences under capital are in part functions of the fact that, lest it fall into chaos and revolution, a system of capitalism must maintain its personnel of workers. Further, both the demands of people themselves for food and health and the dynamics of the sciences have contributed to the explosive growth of biological knowledge during this period. Simultaneously the psychosocial sciences have grown rapidly in response to the accelerated interaction of persons with one another and the growth of large-scale institutions. Thus augmenting and intensifying the tendencies within manufacturing and agricultural industries toward increasing socialized forms of production, the activities of theoretical and practical scientific workers and those associated with them—whose numbers amount to millions—have created a mode of production whose form and quality are reflected in the current consciousness about health and "the ecological crisis." Man as a social and natural being has always faced a crisis with his environment, insofar as food and other biological necessities have not been readily accessible to him. What is qualitatively unique about the present situation is the convergence of population growth with a mode of production of foods and medicines that has brought that population into being and promises in principle to feed and care for the new population. The new sense of "crisis" expresses not only concern about the numbers of mouths to be fed but also hope about the prospect of feeding them. So far as this crisis of the late twentieth century remains, with its antagonism between the socialized, cooperative, scientific mode of production, and the individualized, acquisitive, irrational structure of relations in social institutions, the crisis cannot be *reduced* to the simple categories of class struggle of the nineteenth century, though class struggle in its new forms remains at the core of the human problem.

Ecology is a uniquely modern science, having begun to form only in the last century and to take on the shape of a science in the recent decades of this century. In its farthest reaches, it is a science connecting many sciences, and in this sense is, as we said earlier, a form of dialectics. It is an effort to summarize and interpret the dialectical interconnections of the sciences as they bear on man's nature and destiny, and hence it is the supreme human science. As a science it seeks not only to understand the world but implicitly to change it. It is therefore future-oriented. Comprehensive ecology, humanistic ecology, articulates a hope and vision of what man is in process of becoming, for good as well as ill, and a guide and hope of what he ought to become. Present-day ecology is a reflection of the socializing and naturalizing processes already at work in human society—the processes that bind man to man and man to nature. It reflects man's mode of production, his socialized life and work in relation to nature.

The logic inherent in the ecological study of the "facts" of nature leads one to discover and secure the conditions of a system of nature and society that conduces to the welfare of man. Such a logic, however, derives from coupling to the facts the premise that man's survival and welfare are important. Without that premise, we have no such logic; we have only "bare" facts, denuded of broad humanistic concerns. But who would be content with such a dehumanized approach? Who would study nature and blot out of consideration its reticulations that reach out into the life of man, that ramify there and either brace it or wreck it? Unfortunately, we can find such computerized contemplators who pretend to be "neutral" in ecological questions bearing on man's life and welfare. But the pretense does not hide another "fact"—that the policies of capitalism are rapidly eroding not only the ecology of nature (the "facts" for contemplation) but also the "neutral" contemplators themselves. That anticontemplative fact is a fact that the contemplators choose to ignore.

More to the point, the logic of this humanistic ecology leads to a practical imperative—the imperative of a democratic, socialized society in harmony with nature. Sometimes, otherwise able and concerned ecologists avoid or criticize such an imperative. For example, in *Biology and the Future of Man,* a book described by C. H. Waddington as "by far the best single volume of the whole biology that is now available anywhere," a number of scientists discuss in a rational way the worldwide problems of war, nuclear weapons, population growth, food production, and the degradation of the environment. They refer to such problems as "rising atmospheric CO_2; increasing particulate content of the atmosphere; buildup of radioactivity; accumulation of diverse chemicals in lakes,

streams, rivers, coastal waters and the ocean itself; soil erosion and destruction; replacement of fertile green farm and woodland by highways and
towns; rising noise levels; and "thermal pollution," as well as the use of
nondegradable detergents and pesticides, heavy use of fertilizers, internal
combustion exhaust, sonic boom, and the vapor trails of supersonic trans-
port. They conclude that if the projection (to the year 2000) for expansion of industries influencing pollution is approximately correct—"and if
each of these enterprises grows without appropriate monitoring for its
ecological consequences"—then "the totality could constitute the saddest,
most brutal and disastrous act of vandalism in history."[207] Their conclusion is correct, but they fail to name the principal vandal. And who is to
do the "appropriate monitoring," the "regional and national planning"?
The enterprises themselves? We have no evidence for even a slim assurance
of such monitoring. The government, as demanded by the people, or the
people themselves? In that case, profits must cease to determine economic
and social policy. That means socialism. The scientists drew their conclusion,
but did not let themselves be drawn to its complete end.

Ian L. McHarg, who writes so movingly of nature's designs and who
righteously lambasts the despoilers of the environment, does not, so far
as I can tell, come out on the side of socialism.[208] Similarly, Loren
Eiseley, a poignant scientist and a knowing poet, expresses a typical un-
scientific and unpoetic stereotype of socialism: "The group ethic as distinct from personal ethic is faceless and obscure. It is whatever its leaders
choose it to mean; it destroys the innocent and justifies the act in terms
of the future. In Russia, this has been done on a colossal scale."[209]

Statements of this kind can be multiplied. This is not the place to deal
with all the criticisms directed at socialist states. But I cite this criticism
by Eiseley as typical of a widespread misunderstanding that bears on the
issues of ecology in socialist states. If we take the Soviet Union as an
example, it is necessary to understand the sixty-year development of this
country in all its complexity—its problems, opportunities, failures, and
successes. And here we can only suggest this complexity in a simplified
and general way. The usual Western perspective of the Soviet Union concentrates on Stalinism and tends to see Stalinism as the prime factor in
Soviet life. The errors and evils of Stalinism are now generally admitted
in the Soviet Union: the illegal repression of fundamental rights, the cult
of the authoritarian individual, the violation of democratic and collective
procedures, secret police, terrorism and unjust imprisonment, labor camps,

sterile bureaucracies, and suffocation of creative thought and initiative. Alongside of this we must also take due note of the struggle of the Soviet people to preserve their new state against foreign military intervention, civil war, and famine; the massive tasks of developing socialist industrialization and collective agriculture (led by Stalin); economic hostility and military encirclement by capitalist powers; internal disruption from Trotskyites, Zinovievites, and rightists; esponiage; the rise of Nazi Germany supported by other Western powers; the refusal by these powers to accept the Soviet policy of collective security; the eventual devastating invasion of the Soviet Union by the Nazi armies, killing more than 20 million people; and the rigorous demands of postwar reconstruction, coupled with the cold war. Through all this the people of the Soviet Union eradicated economic classes, mass unemployment, and hunger; raised the level of literacy, medical care, child care, ethnic, racial, and women's equality, old age security, and the general standard of living to a level comparable to the highest level in Europe. No society, however nobly conceived—and we are reminded of our own and other democracies—is free of defects of planning, of implementation of plans, and of human custom and character. The people of the Soviet Union have borne their own share of defects and have created their own unique values.

Soviet ecological policy is illustrative of both the problems and achievements in socialist society. During the 1920s and 1930s, the pressure to build a modernized industrial and agricultural base, and to do that speedily so as to meet the demands of the citizens and to strengthen the society against external attack and destruction, was so intense that considerations of the preservation of environmental balance and quality became secondary. The years of the Great Patriotic War consumed all energies in the overriding task of defense, survival, and the defeat of Hitlerism. After the war, as after the revolution, energies were once more mobilized for reconstructing a devastated country. The "Great Plan for the Transformation of Nature," announced in 1948, was a grand scheme of afforestation, weather control, grassland crop rotation, and water storage. But by 1953 many parts of this plan had been discontinued. Large-scale erosion in the Ukraine became a problem; and the opening up of 100 million acres of virgin lands in the middle 1950s created problems of soil depletion because the scanty rainfall had not been reckoned with.

From 1959 to 1966, a cellulose plant was built on the shore of Lake Baikal, the largest body of fresh water in the world, and the plant discharged its wastes into the lake. During construction, a great public debate raged over the industrial use and pollution of the lake. The upshot was the installation of a filter system that minimized pollution. This controversy was a major factor in alerting the Soviet public, planners, and party officials to the importance of ecological values and of incorporating them into industrial and other kinds of planning. At present writing, ecological care in the Soviet Union is a central concern at state and local levels, and the Soviet Union has renewed its call for an all-European conference on environmental protection, transportation, and energy.

Critics of the Soviet Union, like Marshall I. Goldman, contend that its ecological problems are similar to or "convergent" with those of capitalist countries. The evidence, however, does not bear out this contention. The Soviet Union does have ecological problems—air pollution, some industrial pollution of inland waters, erosion, and so on; and some of these problems have been long-standing. But there is one significant difference between the ecological problems in a socialist country and those in a capitalist country: in a socialist country, the problems are rooted not in a productive system of private ownership and profit but in a system of public ownership, and consequently when the public and the state officials are alerted to a problem and aroused to do something about it, solution comes with relative ease. We have referred to the example of the campaign around the pollution of Lake Baikal. William M. Mandel, who has refuted Goldman's views in the journal *Environment* (December 1972 and May 1973), has elsewhere (in *Science and Society,* Winter 1972) cited a report from the Nature Protection Society in the Soviet Union which indicates a vast reduction in the discharge of industrial untreated effluents into waters as well as a significant increase in water-treatment and gas-and-dust installations in the five-year period ending in 1971. This Society exercised its legal right to demand that officials up to cabinet rank deal with pollution directly and its right to investigate the functioning of industry, agriculture, and the like. Mandel believes that the turning point in the fight for environmental protection in the Soviet Union came in 1966. A major force in this turning point was an awakened public awareness.

To understand the ecological problem not only among states of different social systems but also on a global scale, we must see the systems of capitalist and socialist states in interaction and must grasp the dialectics of that interaction. As the idea of ecology expresses an awareness of the interconnectedness of things in a world transformed by technology, so Marx and Engels, tracing the course of technology in history and in their time, perceived the ecological demands made upon that technology if it were not to continue to be—as it was under capitalism—self-consuming. That is, they saw two modes of social life in conflict—the forces of actual capitalism and the forces of unactualized socialism—attended by their two conflicting modes of action toward nature—myopic and dehumanizing destruction and planned usage and edifying reconstruction. Then capitalism prevailed and reigned; socialism still slumbered in embryo, a thing of slow and steady growth, though in Marx's day it had already begun to kick against the walls of the womb. What led Marx and Engels to believe that the one would be displaced by the other? Not hope or moral conviction but the belief, based on observation of and reflection on the data of human history, that the ecological forces binding persons one to another and to nonhuman nature will be stronger, if organized and exerted, than the forces dividing them. Moreover, Marx and Engels held that the forces of modern technology, which intensified these conditions, were driving men into a new world of unity whether they liked it or not. This driving movement has endured with unabated momentum down to the world of our own day.

How can we characterize the technological features of such a world? Those features are the worldly features of technology. Agriculture, fishing, forestry, mining, manufacturing, transportation, commerce, communication, science, art, politics, religion, and virtually any other kind of technique that we can name have planetary interconnections. Everything is drawn along into the tide moving toward a "world market," everything drives toward internationalizing and universalizing itself. Science by definition looks with unconcern at provincial and national boundaries in its search for comprehension and application of the truth. So far as a technique tends to become scientific, as all techniques do, it tends to transcend boundaries in the same way. Scientific knowledge and technical know-how, once communicated by oral or printed symbols, tend to spread and to universalize themselves as needs demand.

Marx and Engels saw this struggle toward a universal world society—"of the people, by the people, and for the people"—going on throughout human history but repeatedly blocked by the particular limits of a ruling class. For them, every historical social-ecological pattern based on the dominance of one class over another, on the aggression of individual against individual, on group against group and man against nature, and on the hierarchies of power and social role has organized an economy, fed a surplus population, and eventually collapsed under its own inability to organize its productive techniques and resources to sustain its members. The social ecology of aggression and dominance, adapted to the hominid hunters' pursuit of animals on the open savanna[210] and developed in class societies throughout history, has proved dysfunctional for modern urban living, as it was dysfunctional in past urban societies. As an alternative, Marx envisioned a social ecology that dispenses with the dominance-submission pattern, is cooperative, and is organized for the optimal satisfaction of human needs of both infant and adult in the setting of a sustaining nature. Such a new social organization would reproduce the cooperative unity of the mammalian, primate, and primitive hominid group, but at a more complex level. Socialism would be and is a restoration of man to a social form adapted to his nature and to a nature adapted to his social character. The epoch of class ecology must become an interim between primitive communal ecology—"such as we find it at the dawn of human development,"[211] among hunters and agricultural communities—and socialist advanced ecology. This interim is "a definite limited epoch"[212] in the history of society. In a fitting organic figure of Marx, the productive power and socialized labor of the new order, impelled by the dynamic of its superior adaptability, will "burst asunder . . . their capitalist integument."[213] The pattern of cooperation, which already "constitutes the fundamental form of the capitalist mode of production"[214]—though in conflict with the social relations of capitalism—will extend its sway to the whole of society and nature. This social-ecological prediction of Marx is confirmed by the fact that since 1917 the "bursting asunder" has occurred among one-third of the world's population.

Modern industry, of which capitalism is only *one* form, is uniquely dynamic: "all earlier modes of production were essentially conservative."[215] Those modes conserved human relations in the family, on the land, and in the craft shop. They conserved man's static and regular relations to external nature. Capital, however, impinges on man's social and natural relations; it invades, disrupts, and scatters them. What is required, therefore, is the creation of a new kind of equilibrium which unites man, newly liberated from stifling relations, with his fellow man and with external nature in ways that conserve the conditions of his dynamic growth.

Socialism as economy is only the groundwork for a new planetary ecology. It has yet far to go in perfecting itself. As Lenin remarked, "it is much easier to seize power in a revolutionary epoch than to know how to use this power properly."[216] Harmful antagonisms within persons (neuroses, psychoses, alienations), between persons (rivalries, the cult of personality), between groups (suppression of essential freedoms), between socialist nations (chauvinist conflicts), and between nations and nature, all exist in the socialist regions of the world. Among the factors accounting for these antagonisms are social ones but also biological ones of which some Marxists have not taken full measure. Man as species and individual is a product of billions of years of biological evolution. Each person carries within his organism dispositions to respond to the world of nature and social relations in ways that have proved adaptive to the past conditions of his animal and hominid prehistory—the hungry protozoa, the adventurous amphibia, the placental mammals, the brainy mammalian social group, the primate family, and the small hominid band on the open savanna using speech, tools, and fire. Urban history with class society is a very brief period in this long evolution. The dispositions to seek pleasure and avoid pain, to eat, to grasp and suckle, to respond to threatening situations with fight or flight, and to engage in sexual relations are very deep dispositions, coded in the genetic structure and mediated by the primitive parts of the brain.[217] Only through social learning are these dispositions brought under the control of the higher centers and integrated with the uniquely human functions of visual perception and manual control, foresight, social care and intelligent cooperation for the welfare of the group, the species, and the planetary ecological system.

Social Darwinists have assumed that because modern men, like men 500,000 years ago, respond to situations interpreted as dangerous with primitive alarm dispositions, ready for fight or flight, they are therefore "by nature" aggressive or war-making; or that because they have been and are carnivores, they are cannibalistic. But a tendency to respond to certain stimuli is not the same as preordained behavior. And this is the case especially with man, whose biological tendencies, evolved in the past, are subject in their overt, concrete expressions to the inhibitions, enhancements, and modifications of culture, i.e., symbolic learning. Human behavior, even when it is biologically grounded, as in hunger or sex for instance, is not transmitted only by biological forces; it is mediated and passed from one generation to another through culture. At the same time, culture cannot bend human behavior beyond certain broad limits imposed by biological demands without peril to the group. For example, it must provide for individuals a certain minimum amount of food with a minimal frequency, and it must make available to the individual some forms of association and cooperation with others.

While resurgent social Darwinism, based on erroneous inferences from these primitive dispositions, has given aid and comfort to capitalism, its errors should not induce us to deny the evidence for our biological history encoded in our contemporary bodies. Anyone with a *social* theory of man and history may be inclined to overlook biological and other natural factors. Marxism is *materialism* and cannot overlook the body and its natural environment, nor can it, as *dialectical* materialism, overlook the subtle interplay of the human body and its social and natural environment. We should examine the evidence for those bodily dispositions and ask how they can be controlled and integrated into personality and social life in constructive and creative forms. For example, how can we redirect in socially useful ways the disposition to respond with fear and aggressiveness to threatening situations? Again, there is evidence that our long interdependence with nature generated a need to be with it, to touch and handle it, to smell its odors, hear its sounds, and see its colors and designs;[2] and that, losing such life-enhancing contact with it, we lose a part of our deeper selves. Thus we face the dual problem of (1) readapting to the conditions of urban life the dispositions that developed to adapt man to survive in the wild and (2) preserving in man's new social environment the deep needs of his biological nature.

Marx and Engels recognized in the "revolutionary" technological processes of modern industry a social power rising to ascendancy above all else.[219] In its "universal development of productive forces,"[220] it liberates man and his activity toward a kindred kind of development, breaking down all obstacles and all inhibiting relations. But there is one refractory barrier that bars the way. That is capital itself.[221] It generates an ecosystem which, as it drives toward a universal planetary society interpenetrating all of the biosphere, is blocked by a class-constricted system of ownership, distribution, and exchange, a system characterized by the pathologies of poverty, imperialism, war, racism, and political tyranny. This class system denies the demands of nature's ecological system for dialectical balance of harmony and collision as well as the demands of individual men for fulfillment in cooperation and struggle with social forces and environing nature. It wastes natural resources and is engaged in "the most extravagant waste of individual development."[222] Yet, as Marx and Engels reasoned, capital is also the unacknowledged vehicle of this same development to which it is the principal obstruction and must eventually give way to it in the form of socialism. In their early years, Marx and Engels supposed that this transition would take place in the countries of the world more or less at the same time. Later, they realized that it would come country by country.

Within a generation after Engels died (twenty-two years), this change began to occur. Socialism was consummated in one country, the Soviet Union, which for sixty years found itself the target of protracted antagonism from another system. Despite this antagonism between the old ecosystem of capitalism and the new ecosystem of socialism, socialism has been embraced by thirteen other countries. This set of socialist nations now includes one-third of mankind. The quantitative changes they have wrought have been accompanied by qualitative changes in the character of human societies. Of course these socialist nations exhibit economic, political, cultural, historical, and other differences. Their ecological histories, problems, policies, failures, and successes vary. While bitter controversies of principle and policy now prevail between some of these socialist states, particularly between the People's Republic of China and the Soviet Union, they all collectively constitute an important new socioecological order on the world scene in the last thirty-five years. Each socialist state possesses and wields collective, public control over the environment with the power to transform it in rational, humanistic ways. In principle no person or group stands to "profit" from ecological malfeasance. The natural and social sciences are united in long-range social planning aimed at the systemic development of persons, society, and nature. Planning necessarily includes scientific environmental research whose application is closely coordinated with economic and social planning.

Some people are inclined to overlook the close interconnection between society and nature in the socialist nations, i.e., between securing certain human values, like food, housing, medical care, availability and equality of vocational and cultural opportunity, and other values, and making wise use of the substances and energies in the natural environment. Socioeconomic organization and ecology are not separable. Human need and the actual or potential goods of nature that fulfill such need depend on each other. Socioeconomic policy can destroy or conserve the resources of nature, and nature in turn can destroy or conserve man. Thus "ecology" in the broad and accurate sense must be taken to embrace human social systems, both national and international, as they interact with the natural environment; and "society" properly understood must be seen as inter-penetrating, being transformed by and transforming the natural environment. Social justice is in part a function of how a society treats nature, and sound or "just" treatment of nature is in part a function of the justice within society. In the language of the sciences, the whole global system is an interacting system of energies, i.e., the energies of human societies interacting with the energies of the natural environment.

The emergence of the new ecosystem of socialist nations on the planet represents a new stage in the history of life with respect to the production, management, and use of energy. The first cells and the first urban civilizations with their technologies marked dramatic advances in the production, conservation, and distribution of energy. Then, with its extraction of energy from peat, coal, steam, electricity, petroleum, and natural gas, capitalism vastly accelerated the processes of the capture and conversion of energy. Today it is beginning to make use of the energy in fissionable materials and to experiment with methods for using solar energy.[223] This increase of productive power under capitalism has brought with it many social problems, as we have indicated. By transferring the owner-ship and control of the instruments of production into the hands of society at large, socialist states have with more or less success solved many of these problems: hunger, lack of medical care, illiteracy, unemployment, and so forth. Much of course remains to be done.

It is generally recognized in the socialist countries that ecological progress requires the cooperation of socialist countries with each other. The political principle of proletarian internationalism as well as the facts of modern ecology dictate such cooperation. Similarly it is increasingly recognized in many states of different economic systems that while the emergent ecosystem of socialist nations is antagonistic to the established capitalist ecosystem, the two competing and interlocked systems must cooperate if we are to avoid nuclear or ecological disaster for the whole of our race. Capitalist and socialist societies are interdependent in both the world economy and the world ecology, which they themselves interpenetrate.

Marx and Engels, of course, could not foresee the current relations between the capitalist and the socialist blocs, with their antagonistic and interdependent ecosystems. As early as 1917, however, Lenin grasped that in order for the new socialist system to survive it had to achieve peaceful coexistence with the other.[224] First of all, such a policy had to be based on trade and durable economic relations between both sides.

Second, it had to include the mutual renunciation of war; the use of negotiation in international disputes; mutual respect for the sovereignty, independence, and equality of states; and scientific, technical, and cultural exchange. The point of peaceful coexistence then as now is not only to prevent the outbreak of wars harmful to both sides but also to secure "mutual benefits" to both sides. As atomic, chemical, and bacteriological weapons developed after World War II, the point of avoiding mutual harm and securing mutual benefits became more urgent than ever, and the agreements between the United States and the Soviet Union in 1972 and 1973 were big strides toward making concrete the policy of peaceful coexistence.

Lenin recognized that it was necessary to establish socialism on a sound economic footing in one country as the first step toward transforming all states into socialist ones. In order to effect the spread of socialism, a foreign policy of "good-neighborly relations"[225] was necessary. The construction of socialism in one country was the prime task, and that required trade with other (capitalist) countries. After World War II and the rapid growth of socialist states, peaceful coexistence remained a goal for these states, though its attainment was slowed by the cold war. Nonetheless, technological advances in the past thirty years on both sides have created such worldwide problems that the two systems increasingly find themselves confronting a common economic-ecological environment, though in different ways and with different goals in view. Their recognition of this common environment with its common problems has been evidenced in recent years in the agreements between the governments of the United States and the Soviet Union affecting their common planetary environment, both social and natural: the International Geophysical Year (1957-1958); the agreement on cultural and scientific exchange (1958); the treaty demilitarizing the Antarctic (1959); the treaty banning nuclear weapon tests in the atmosphere, outer space, and under water (1963); the International Years of the Quiet Sun (1964-1965); the treaty on the nonproliferation of nuclear weapons (approved by the U.N. General Assembly, 1968); the treaty to limit ABM's (1972); the interim agreement to limit strategic offensive weapons (1972); the agreements on co-operation in the fields of medical science, public health, science and technology, the exploration and use of outer space for peaceful purposes, and environmental protection (1972); the agreement on prevention of incidents on and over the sea (1972); the agreement to prevent nuclear war (1973); the agreement on the limitation of strategic offensive arms (1973); the agreement to increase commercial and economic ties between the two countries (1973); and agreements for bilateral cooperation in the peaceful uses of atomic energy, in agriculture and world ocean studies, in transportation, and in contacts and exchanges (1973).

That both capitalist and socialist states acknowledge these common environmental problems is an aspect of the worldwide socialization process that Marx described and that is at work today. Ecological truths are planetary truths. So are the truths about atomic fission and fusion, atomic warfare, mass death, and genocide. Any general law pertaining to the behavior of things or people obtains in Washington, Moscow, Peking, Tokyo, and Calcutta. But ecological truths are inclusive and integrative of many such laws, and pertain not only to local, national, or continental conditions but also to the planet taken as a whole. Thus, the common concern of governments with ecological conditions and truths, and the increasing sharing of such truths, are reflections on a conscious level of the gathering worldwide movement toward a unitary planetary society such as Marx foresaw. The economic and political shape of this new society, he thought, would be socialist. Today, some charge that the new detente between capitalist and socialist states is a sellout to Marxist-Leninist socialism and its revolutionary thrust and that in the cooperation of the two kinds of states the socialist states will compromise with or converge toward capitalist states. Such criticism overlooks the dialectics of detente. In that dialectics, ecological truths, and the truths about the dangers and needed controls of modern weaponry, can be confirmed and developed in both theory and cooperative practice. These are truths which socialism in general principle and practice has upheld, truths which are organically related to the humanistic social organization of socialist states. By the affirmation and application of these truths, the whole impact of socialist systems will make itself felt in government circles and in the media. At the same time, the claims of capitalism to be free from public demands, to put profit ahead of human welfare, to pollute the environment as it pleases, and to conduct aggressive war without check must be rejected. In the dialectics of cooperation with socialist states, capitalist states must modify their policies and ultimately be transformed. Of course, there will be setbacks and covert and overt antagonisms, but the general tendency will force capitalism to adapt its practices to the worldwide socializing tendencies.

The discussions in recent years between the governments of the United

States, the Soviet Union, and the People's Republic of China are signs of
a growing recognition of common ecological problems that transcend
social systems and require international solution. We have mentioned the
1972 and 1973 agreements between the United States and the U.S.S.R.
which pertain to the prevention of nuclear war, the limitation of strategic
armaments, the peaceful uses of atomic energy, commercial and economic
relations, environmental protection, public health and medicine, outer
space, science and technology, agriculture, world ocean studies, transportation, and cultural and scientific exchanges. While the agreements are only
a small beginning and must eventually involve all nations on a basis of
mutual respect and equality, they signify the forces of science and technology working "behind the backs" of nations to create a planetary
ecology.[226] These forces of such scope and speed have introduced into
human society a new character and quality, a new state of contradiction
between the forces of ecological death and the forces of ecological life.
This contradiction cuts across the opposition between capitalist and
socialist states because, while the ecological crisis is less severe in socialist
than in capitalist countries, the ecosystems of both types affect one another and are affected by the inclusive planetary environment. Both
together suffer from ecological distress and death, and *together* will be
saved from them. The nuclear arms race after World War II, for example,
produced and is still producing, by both capitalist and socialist states,
atmospheric and underground "testing" of nuclear devices which, as
scientists throughout the world agree, result in thousands or perhaps
millions of individual deaths and genetic damage for countless generations.[227] The fact that the transformation of capitalism into socialism
is a necessary step in the salvation of mankind is not an excuse for doing
nothing now but rather a compelling reason for doing whatever can be
done to reverse ecological death.

The understanding and solution of this ecological contradiction cannot
be reduced to an oversimple doctrine of class struggle, the classical Marxist
doctrine of wars against capitalism, or the revived Trotskyite theory of
worldwide, indiscriminate, violent overthrow of capitalism. The dangers
of modern nuclear, biological, and chemical weapons,[228] the pollution
of the planet's land, water, and air, the contamination of food chains,
the desecration of wildlife and natural beauty affect *all* persons, classes,
and nations, and if they are not solved soon the planet will not be a fit
place for any person, class, nation, or ecosystem. In their cynicism,

capitalists have never seriously claimed to create a world of life, health, peace, and happiness for the masses. But socialists, who earnestly aspire to this kind of world, will fail if their world becomes one of death, illness, war, and unhappiness.

We have already observed that poor and developing regions and nations dominated by the need to survive may not give much attention or energy to problems of ecology. This fact itself is indicative of the direct connection between an improved economy and an improved ecology. It points to the need for genuine economic aid to such countries by the more developed ones. Such aid, under the conditions of mutual respect for the sovereignty, independence, and equality of all nations in the development of their natural and human resources in their own ways, has already been given by some nations. But since, as U Thant and others have pointed out, the gap between rich and poor nations is daily widening, the aid must be stepped up, broadened, and made systematic. In 1973 the Soviet Union proposed to the United Nations a 10 percent reduction in the military budgets of the five permanent Security Council members and a use of part of the saved funds to help developing countries. International development and its concomitant constructive ecological policy depend on detente, peaceful coexistence, and disarmament.

Developed capitalist nations are beleaguered by ecological problems. At the same time ecological consciousness and movements in such countries have greatly increased in the past decade; environmentalist groups in the United States have scored notable victories in legislative bodies and the courts; trade unions have become more active in ecological causes; and the struggle to improve the environment where we live, work, play, and enjoy our recreation is gathering strength. Like the peace movement in the United States, the ecological movement is growing but is divided into many diverse groups that are not yet united by any overall political tactics or economic or ecological philosophy. American ecological theorists, moreover, are frequently pessimistic about the ecological future of mankind. They are inclined to universalize their own national conditions and to overlook, misunderstand, or misrepresent the concrete ecological accomplishments of socialist societies.[229] But we can expect that as theory and practice are related dialectically to one another in the American context of solving economic and ecological problems, the pessimism will give way to a realistic optimism. As people who work collectively in their neighborhoods, regions, states, and nation succeed in making their voice and will known and have their way about ecological policy, their understanding of the basic causes of economic, ecological, and social problems will simultaneously deepen. They will confront and seriously consider the great alternatives of private appropriation and social ownership, of capitalism and socialism. And that consideration will lead to a balanced assessment of what has been and is being done in past and present socialist economies and to a realistic hope of making basic changes in the economic and social structure at home.

What can we say of the ecology in socialist economies? In this work we have not taken as our task a thorough investigation of the pros and cons of ecological policy and practice in these economies. The record is mixed. But all the socialist economies in recent years have given new attention to ecological issues, and once launched on a policy of ecological transformation they have been strikingly successful. In the Soviet Union there is increasing concern for "preserving the dialectical balance in the biology of organisms and all living things,"[230] as well as efforts to "preserve and beautify" the land "for present and future generations of Soviet people."[231] In the People's Republic of China, much work has gone into the control of environmental pollution, especially the disposal of industrial waste and urban sewage.[232] Much has been accomplished,[233] but much remains to be done.

It has been argued that ecological problems will be solved only when nations become socialist. The answer: (1) two-thirds of the world's people are not yet socialist, and the world's ecological crisis is deepening; (2) existing socialist states have many ecological problems; (3) ecological problems are planetwide, spreading from nation to nation; (4) socialist persons and nations have a responsibility to all the world's peoples, both now and in the future, and hence a responsibility for the creation of a healthy planetary environment for the present and future, because a victory for world socialism in a world of empty nature is an empty victory. The task for the present-day socialist is not merely to move his nation into socialism or to improve its established socialism. On the principle of proletarian internationalism, it is also to help create a planetary environment required by and worthy of man under socialism. The pre-World War II formula, socialism by nation and then a better environment, must be supplemented by the late twentieth-century formula, ecological solutions as dictated by the demands of coming world socialism. Such solutions cannot be left to governments alone but must be achieved by peoples cooperating across national boundaries.

Peaceful coexistence, including ecological cooperation, does not mean that the class struggle and the ideological struggle between the two eco-systems of capitalism and socialism will not and ought not to continue. It does not negate but rather confirms the truth that Marxism has always proclaimed: that the social ecosystem that is most balanced, most dynamic, most viable, most responsive to both the needs of the people (humanistic) and the laws of nature (scientific) is the ecosystem that will prevail over all others. As we move toward a truly humanized planetary environment, we will begin to realize concretely what Marx meant by man's "human science." In such a science, concretely understood, man's activity is a natural relation of nature to itself, a mutually confirming relation that reveals man's natural powers. It is also a humanizing of both man's nature and that larger nature from which he comes, on which he depends, and whose destiny is now more than ever before indissolubly linked with his own.

The old ethical imperative, "Cooperate or perish," is today more urgent than ever. An escalation in the arms race and in bacterial, chemical, and nuclear weaponry, as well as a spread of ecological contamination, will, if not checked and reversed, result in irreparable damage to our race and our environment and the eventual genocide of our race. The saving alternative to this fatal course is international cooperation aimed at the survival of nations and peoples and care of the one environment that we all share. People in their very nature, in their life and fulfillment, are united with one another and with their natural global habitat; people taken collectively and their planetary environment are therefore destroyed or saved together.

The imperative of international cooperation requires cooperation among states through their governments and other state agencies. To this end, the Final Act of the Conference on Security and Cooperation in Europe, signed at the Helsinki Summit Meeting on August 1, 1975, by thirty-three nations of Europe and the United States and Canada, was a vital step and a model for international cooperation in creating zones of peace and collective security throughout the world.

Equally important are the cooperative actions of the peoples themselve the working people of the world. The class struggle against dehumanizatio and oppression has been a fact of human history since the beginning of history. Today the struggle of the working people is called upon not only to maintain socialist societies wherever they exist and to create them where they do not exist, but also to preserve and improve the planetary environment against the disruption and impairment of nonhuman nature imposed by the capitalist mode of production and exchange. This struggle against environmental disruption cannot be confined to a single nation, region, continent, or party, or even to many of these short of the whole globe of humankind. The law of the sea, the law of the land, the law of the air—in short, the laws of nature—have no boundaries. In the *Manifesto of the Communist Party,* published in 1848, Marx and Engels, in response to the criticism that the workingmen desire to abolish countries and nationality, declared that "the workingmen have no country." They immediately added that "the proletariat must first of all acquire political supremacy" as the leading class of their given nation in order to achieve worldwide power. They thought that international communism was not far distant. Today, the working class struggle for the political transformation of individual nations into communist nations still continues.

Whereas the political struggle since 1848 has been a struggle for *social* emancipation, it is now clearly a struggle for humanity's emancipation not only from *social* but also from *natural* oppression. The struggle for emancipation from environmental oppression cannot wait for the political emancipation of a given nation or group of nations. The ecological struggle must be linked to the political struggle. It must be made clear that the class struggle against political oppression of the workers by a ruling class is simultaneously a struggle against the oppression of nature by that same ruling class, the ruination of natural sites of recreation and beauty, and other forms of environmental degradation. Liberation from one entails liberation from the other, since human beings are an integral part of their environment, inextricably bound to it. Just as, under capitalism, a very small ruling class imposes its will and values on the rest of humanity and nature, so a very large working class on national and international levels can depose that imposition and liberate our human-natural-ecological system from the devastation to which it has been subjected for so long.

The ecological movement in capitalist countries, which is so frequently dominated by well-intentioned liberals who are not aware of the factor of class in society, must be informed with a class perspective and infused with class action. On the other hand, left-wing political movements, sometimes working from a narrow nineteenth-century emphasis on class struggle in the industrial trade unions, must broaden their struggle to embrace the ecological one. On the international plane, all of us ought to work with all groups sincerely aiming at ecological improvement, just as we ought to work with all groups aiming at peace. The reason is simple: the

survival of the human race, whether living in capitalist or socialist societies, depends on the prevention of our quick death by armaments and
war and our slow death by ecocide.

Workers in bondage to long hours, low pay, a rising cost of living, and
degradation of morale on or off the job (condemned to the punitive
indignity of unemployment) are also in bondage to unsafe conditions at
their workplace, pollution at work and at home, the wastage of natural
resources, and countless other forms of environmental destruction. As
could be seen in the aggressive wars in Korea and Vietnam, imperialism
is no respecter of boundaries: in its all-consuming urge for investments,
sales, and profits, it is blind to the consequences it wreaks on the social
and natural world. It is indifferent to the fact that war means destroying
villages, land, vegetation, water, and culture; that private "investment"
in a developing country means tearing up once more the natural and
cultural ecology; that increased consumption at home means depletion
of resources, litter, junkyards, and the cheapening of popular taste; and
that a cancerous national budget for military expenditures and civilian
government means multiplying parasitic unproductive classes at the expens
of both society and nature.

Where will it all end? There is nothing automatic and inevitable about
the success of the class struggle. The movement toward a spiritual transfor
mation of humanity—developing along with and subsequent to the transition from capitalist ecology to socialist ecology—is not a vague movement
of the *Weltgeist*. It is a movement composed of specific people in specific
neighborhoods working in specific ways to create a better society and a
better ecology simultaneously. The responsibility and the opportunity
for this better human world belong to us, the working people of the
world—the ones who really count if they stand up to be counted in the
struggle to make a more humanized nature and a more naturalized human

Notes

1. Joseph Needham, *Order and Life* (Cambridge, Mass.: M.I.T. Press, 1968),
p. 108. Needham is paraphrasing C. H. Waddington. A field is a system of ordered
individuals. In my discussion\an "individual" is defined as a discernible unit of
energy producing bounded, determinate causes and undergoing bounded, determinate effects in relation to its environment. An individual may be a participant

in the system of another individual, and a system normally contains subindividuals within its own internal system. The difference between a group of individuals forming a superindividual and a group not forming a superindividual is a matter of degree. The reason is that the term "definite" in Needham's definition is arbitrary.

2. For this definition of a living system, see J. D. Bernal, *The Physical Basis of Life* (London: Routledge and Kegan Paul, 1951), pp. 23-24.

3. Philip Handler, ed., *Biology and the Future of Man* (New York: Oxford University Press, 1970), pp. 431-434.

4. Arthur O. Lovejoy, *The Great Chain of Being* (Cambridge, Mass.: Harvard University Press, 1964), p. 52. Theologians like Leibniz and Paley emphasized the divine design of nature, but Voltaire and Hume showed what a strain this view placed on reason. Pascal, who came before them, contrasted rational man with the immensities of space, insensate matter, and instinctive animals. He saw man in an alien universe where even "a vapor, a drop of water suffices to kill him." Here we have two extremes. Neither position took account of the web of living relations woven by plants and animals on the planet.

5. E. J. Hobsbawm, *The Age of Revolution 1789-1848* (New York: New American Library, 1962), p. 346.

6. Cf. S. Alexander, *Beauty and Other Forms of Value* (New York: Thomas Y. Crowell, 1968), ch. 17, "Sub-Human Value; and Value in General." "Moisture is a value for plants because it satisfies an organic need" (p. 288). A similar view is held by a biologist, C. Judson Herrick, in *The Evolution of Human Nature* (Austin: University of Texas Press, 1956), ch. 12.

7. As evidence that this term is not merely a quaint nineteenth-century organismic metaphor applied to social life, see Abel Wolman, "The Metabolism of Cities," *Scientific American* 213, no. 3 (September 1965): 179-190.

8. That Marx considered the influence of natural conditions on production to be important is indicated by his reference to the "epoch-making" work of J. Massie on the subject. Karl Marx, *Capital*, vol. 1, ed. Frederick Engels (New York: International Publishers, 1967), p. 514 n. 3. For a recent view, cf. Ellsworth Huntington, *Mainsprings of Civilization* (New York: John Wiley and Sons, 1945), Ch. 20.

9. Karl Marx and Friedrich Engels, *The German Ideology*, pts. 1 and 3, ed. R. Pascal (New York: International Publishers, 1947), p. 57.

10. Ibid., p. 81.

11. Darwin in turn likened natural selection to man's selection of plants and animals: "If man can by patience select variations useful to him, why, under changing and complex conditions of life, should not variations useful to nature's living products often arise, and be preserved or selected?" Charles Darwin, *The Origin of Species, by Means of Natural Selection*, reprinted from 6th London edition (New York: Peter Eckler, n.d.), p. 456.

12. Karl Marx, *Grundrisse. Foundations of the Critique of Political Economy* (Rough Draft), trans. Martin Nicolaus (Harmondsworth, England: Penguin Books, 1973), p. 143.

13. Cf. the pessimistic statement by the natural scientist Hudson Hoagland: "Man's great neocortex may turn out to be a phylogenetic tumor capable of pro-

ducing modern nuclear armaments, but unable to control their use." *The Humanist* 33, no. 1 (January/February 1973): 17.

14. Barry Commoner, *Alliance for Survival* (New York: United Electrical, Radio and Machine Workers of America, 1972), pp. 16 ff.

15. Gus Hall, *Ecology: Can We Survive Under Capitalism?* (New York: International Publishers, 1972), p. 7. If it is objected, as it often is, that merely to displace the corporations with popular control is to democratize the greed of men, we answer: the great majority of people now are not greedy for surplus-value; people's attitudes are shaped by the social system under which they live; and eradication of the corporations, requiring a mass effort, would *ipso facto* carry with it a revolutionary change in the moral sensitivities of most people toward other persons and nature.

16. Marx, *Grundrisse*, p. 410.

17. Karl Marx and Friedrich Engels, *Manifesto of the Communist Party*, ed. Friedrich Engels (New York: International Publishers, 1948), p. 14.

18. Ibid.

19. Ibid., p. 15.

20. *The German Ideology*, p. 57.

21. *Capital*, vol. 1, pp. 732-733.

22. Ibid., pp. 505-507. See also p. 604.

23. Karl Marx, "On the Jewish Question," in Karl Marx, *Early Writings*, trans. and ed. T. B. Bottomore (New York: McGraw-Hill Book Co., 1964), p. 37.

24. *Capital*, vol. 3, p. 88.

25. Ibid., p. 260.

26. See also ibid., p. 603. Marx speaks of "a free gift of Nature's productive power to labour" in referring to natural elements that pass into the productive process as agencies (*Capital*, vol. 3, p. 745). Capital gets such "labor" free just as it gets human labor free. That is why capitalist countries have an energy (labor) crisis. Who owns the waterpower, atomic power, soil, ores, oil, and human laboring bodies of the planet? Who owns the ecological system?

27. Marx did not use the expression "the laws of human needs," but he presupposes the idea.

28. *Capital*, vol. 3, pp. 121, 614-615.

29. Ibid., p. 121.

30. The recurring reference of Marx and Engels to the problem of human waste in the cities was not a Freudian preoccupation. It sprang from the objective fact of fetid human manure and other solid waste piled up for months or even as long as a year in the courtyards and alleys of crowded working-class slums. The putrefaction flowed down into dank basement dwellings where tens of thousands lived, and its noxious gases filled the air both outside and inside the drafty tenements, nine of whose ten rooms were windowless. See Annette T. Rubinstein, *The Great Tradition in English Literature from Shakespeare to Shaw* (New York: Citadel Press, 1953), p. 636. It was an ecological mess and a living, unrelieved hell for millions of workers and their families. But what did the factory and tenement owners care about such stench and debasement in which men rotted away and manure accumulated for the profit of those who hauled it off? No more than they now care about the slums of

New York and Calcutta or the permanent radioactive befouling of the gene pool of the human race.

Engels graphically described the living conditions of London's poor in 1844: stifling gaseous air, "the filth and stagnant pools" where the working people lived, poorly ventilated dwellings, polluted rivers and drinking water, the accumulation of garbage and offal in the streets, "dozens in single rooms," damp cellar dens, leaky garrets, and "rotten clothing." These two and a half million poor, hungry and diseased, were "hunted like game," were worked daily to exhaustion, and were so deprived that they enjoyed only sex and drink. This was where "ecology" hit home in the nineteenth century, and where it still does today for the vast majority of the two billion people in the capitalist world. (See Selection IX C.)

31. Mentioned by Engels in his letter to P. L. Lavrov, November 12-17, 1875. Karl Marx and Frederick Engels, *Selected Correspondence* (Moscow: Foreign Languages Publishing House, n. d.), p. 367.

32. Friedrich Engels, *Dialectics of Nature* (Moscow: Foreign Languages Publishing House, 1954), p. 58.

33. Population is a clue to food available. In 1650, the world population was 500 million. In 1850, it had doubled. By 1930, it had doubled again. (By 1975, it doubled again, to 4 billion.) Philip Handler, op. cit., p. 902.

34. *Capital,* vol. 1, p. 643.

35. Handler, op. cit., pp. 434, 435.

36. Engels, *Dialectics of Nature,* p. 27.

37. William L. Thomas, Jr., ed., *Man's Role in Changing the Face of the Earth,* vol. 1 (Chicago: University of Chicago Press, 1956), pp. xxviii-xxxi.

38. Lynn White, Jr., *Science* 155, no. 3767 (March 10, 1967): 1203-1207.

39. Jacob Bronowski, "Technology and Culture in Evolution," *The American Scholar* 41, no. 2 (Spring 1972): 197-211.

40. Marx, *Grundrisse,* pp. 540, 694.

41. In his dissenting opinion in *Sierra Club v. Morton,* Supreme Court of the United States, No. 70-34, April 19, 1972, in which he challenged the plans of Walt Disney Productions to "develop" and "improve" Mineral King Valley in the Sierras for resort and commercial use, Supreme Court Justice William O. Douglas argued that "valleys, alpine meadows, rivers, lakes, estuaries, beaches, ridges, groves of trees, swampland, or even air that feels the destructive pressures of modern technology and modern life" should have the right "to sue for their own preservation." His argument is simple. That valleys themselves do not walk into court and plead their cause is assumed. But "those who have that intimate relation with the inanimate object about to be injured, polluted, or otherwise despoiled are its legitimate spokesmen." Such a relation links the claim to be and to become, on the part of both inanimate and animate natural objects, to the claim of human beings to be nourished by them and to cherish and enjoy them. A river, for example, says Justice Douglas, is the symbol of the life it sustains—fish, insects, otters, elk, and man—and as "the river as plaintiff speaks for the ecological unit of life that is part of it," so the people who are a part of that unit can authentically defend that claim. Inanimate ships and corporations involved in personal interests, Douglas says, have been treated as legal personalities; so should valleys. We add that if nature is man's "own real

body," as Marx maintained, and man possesses a right to defend his own immediate enskinned body against harm and destruction, then he has a right to defend that wider body of nature, including its living forms. The two belong together; the defense of one implies the defense of the other. See Clarence Moore, "The Rights and Duties of Beasts and Trees: A Law Teacher's Essay for Landscape Architects," *Journal of Legal Education* 17 (1964): 185-192; and Christopher D. Stone, "Should Trees Have Standing? Toward Legal Rights for Natural Objects," *Southern California Law Review* 45, no. 2 (Spring 1972). Some argue that the people should physically seize and organize for their own interests the natural riches of the environment. But (1) most people are not ready for that, (2) the police power is in the hands of the industrialists who are exploiting places like Mineral King and could put down with court or bullet any popular seizure less than massive, and (3) the establishment of the *argument* of natural and human rights is now an important step toward the people's eventual ownership of nature.

42. William Wordsworth, "Robb Roy."

43. See Commoner, op. cit., for his own formulation of these principles. He has a fourth "law of ecology," namely, "Nature knows best." But when this law is analyzed and qualified, it seems to come down to: "The planetary ecological system is old and complex and can't be meddled with without harming it," or "Nature's general system will prevail over man's excessive interference." His law does *not* mean, I take it, that this is the best of all possible worlds, that we should not try to control disease and similar adversities for man, or that *all* human interventions in the environment are harmful.

44. Garrett Hardin, "The Cybernetics of Competition: A Biologist's View of Society," in *The Subversive Science: Essays Toward an Ecology of Man,* ed. Paul Shepard and Daniel McKinley (Boston: Houghton Mifflin Co., 1969), p. 292.

45. F. Engels, *Ludwig Feuerbach and the End of Classical German Philosophy* (Moscow: Foreign Languages Publishing House, 1950), p. 37.

46. V. I. Lenin, *Materialism and Empirio-Criticism* (Moscow: Foreign Languages Publishing House, 1952), p. 259. This kind of revision, Lenin is at pains to point out, is *not* the pseudorevision of relativism, opportunism, and reaction, and *not* the "revisionism" which in claiming to update Marxism *downdates* it to be used for partisan bourgeois purposes. Lenin thus scores the Machians "for their purely *revisionist* trick of betraying the *essence* of materialism under the guise of criticizing its form." Ibid., p. 260. It is time now for one or more forms of revisionist "Marxian ecology" to appear, if indeed one has not already appeared.

47. *Capital,* vol. 3, p. 262.

48. Ibid., p. 264.

49. Ibid., p. 250.

50. Ibid., p. 249.

51. Ibid., p. 247.

52. Ibid., p. 256.

53. Ibid., p. 253.

54. Ibid., pp. 232-240.

55. Hardin, op. cit., pp. 281-285.

56. Garrett Hardin, *Nature and Man's Fate* (New York: New American Library of World Literature, 1961), p. 80. The formulation of the principle here is my own.

57. *Capital,* vol. 3, p. 259.

58. Ibid., p. 264.

59. Ibid., p. 44.

60. Ibid., p. 250.

61. Ibid.

62. *Capital,* vol. 1, p. 332.

63. Ibid., p. 330.

64. Ibid., pp. 330-332.

65. Engels, *Ludwig Feuerbach and the End of Classical German Philosophy,* p. 54. Thus, on a broad scale, national crime has been the source of the world market. So says Karl Marx, *Theories of Surplus-Value,* part 1 (Moscow: Progress Publishers, 1963), p. 388.

66. Karl Marx, *Pre-Capitalist Economic Formations,* ed. E. J. Hobsbawm and trans. Jack Cohen (New York: International Publishers, 1965), p. 96.

67. The author of Revelation (6:8, New English Bible) says that the rider of the "sickly pale" horse, whose name was Death, one among the four horsemen of the Apocalypse, had "the right to kill by sword and by famine, by pestilence and wild beasts." This statement reflects the class violence of men with men as well as the violent destructiveness of nature toward man in the declining Roman Empire.

68. *Capital,* vol. 1, p. 233.

69. Marx, *Grundrisse,* p. 694.

70. Ibid., p. 307.

71. Introduction to Karl Marx, *Wage-Labour and Capital* (New York: International Publishers, 1933), p. 12.

72. Ibid., p. 31.

73. Marx, *Grundrisse,* p. 694.

74. Ibid., p. 708.

75. *Outlines of a Critique of Political Economy,* in Karl Marx, *The Economic and Philosophic Manuscripts of 1844,* trans. Martin Milligan, ed. Dirk Struik (New York: International Publishers, 1964), pp. 218-219.

76. "40 Million Child-Workers in the World Today," *The UNESCO Courier,* October 1973, pp. 6-11.

77. Clarence J. Glacken, "Man Against Nature: An Outmoded Concept," in *The Environmental Crisis,* ed. Harold W. Helfrich, Jr. (New Haven, Conn.: Yale University Press, 1970), pp. 138-139. See also Ladis K. D. Kristof, "On the Concept of Conquest of Nature," in 21st International Geographical Congress, India, 1968, *Selected Papers,* vol. 3, *Population and Settlement, Geography, Political and Historical Geography* (Calcutta: National Committee for Geography, 1971), pp. 408-413.

78. Marx and Engels, *The German Ideology,* pp. 7, 19, 31, 39, 41, 42.

79. Clarence J. Glacken, "Changing Ideas of the Habitable World," in Thomas, *Man's Role in Changing the Face of the Earth,* vol. 1, pp. 75-79.

80. Andrew Carnegie, "Wealth," *North American Review* 148 (June 1889): 653-664.

81. Engels, *Dialectics of Nature,* p. 261.

82. Leo Marx, *The Machine in the Garden: Technology and the Pastoral Ideal* (New York: Oxford University Press, 1964).

83. Roderick Nash, *Wilderness and the American Mind* (New Haven, Conn.:

Yale University Press, 1967), ch. 2. Yet, the settlers of the North American continent had a "deep affection" for the landscape they created, and a full account of their history must describe both the dialectical struggle of aggressiveness and affection within individuals and among classes. The struggle continues today, in the form of popular movements defending a healthy ecology against predatory commercial interests. For a brief vignette of such a history, see John B. Jackson, "Ghosts at the Door," in *The Subversive Science*, pp. 158-168.

84. Frank E. Manuel, "Toward a Psychological History of Utopia: From Melancholy to Nostalgia," in *Utopias and Utopian Thought*, ed. Frank E. Manuel (Boston: Beacon Press, 1967), pp. 69-98.

85. Marx, *Grundrisse*, p. 539.

86. Michel Foucault, *The Order of Things* (New York: Pantheon Books, 1970).

87. R. W. Southern, *Medieval Humanism and Other Studies* (New York: Harper and Row, 1970).

88. A. N. Whitehead, *Science and the Modern World* (New York: Macmillan Co., 1929), ch. 2.

89. For a similar view on the antinatural influence of Christianity, see Lynn White, Jr., op. cit. See also A. D. White, *A History of the Warfare of Science with Theology in Christendom*, vol. 1, pt. 2 (New York: Dover, 1960), ch. 1. The early Christian Fathers, unlike Pythagoras, Empedocles, and Plutarch, inculcated no duties toward animals and viewed the human race as "isolated, by the scheme of redemption, more than ever from all other races." Yet, the popular legends about the Christian saints were replete with stories of how animals helped the saints, who reciprocated the sympathy. See W. E. H. Lecky, *History of European Morals from Augustus to Charlemagne*, pt. 2 (London: Watts and Company, 1946), pp. 71-73. Probably there was a class basis for both these dualistic and unitary views. The parasitic members of the church hierarchy were shut off from nature, whereas the common people were forced to make a living by dealing directly with both domestic and wild animals. In theory the otherworldliness of Christianity was qualified by the idea of the continuity of all creatures in the great chain of being. Thus Lynn White proposes Francis of Assisi "as patron saint for ecologists," since he believed in "a democracy of all God's creatures." Saint Francis, let it be remembered, was also a religious communist, and his ecology and communism were not accidentally related but naturally and logically required one another. Communism for him was the human counterpart of the democracy of the creatures. Will it be the same for modern ecologists?

90. Karl Marx, *The Economic and Philosophic Manuscripts of 1844*, p. 114.

91. Ibid., p. 74.

92. Ibid., pp. 156-157.

93. Engels, *Dialectics of Nature*, p. 262.

94. Gustav Mayer, *Friedrich Engels: A Biography* (New York: Howard Fertig, 1969), p. 201.

95. Wilhelm Liebknecht, *Karl Marx: Biographical Memoirs*, trans. Ernest Untermann (Chicago: Charles H. Kerr, 1901), pp. 135-137.

96. Ibid., p. 409.

97. Lynton Keith Caldwell, *Environment: A Challenge to Modern Society* (Garden City, N.Y.: Doubleday, 1971), p. 212.

98. Ibid., p. 103.

99. Karl Marx, *A Contribution to the Critique of Political Economy* (Chicago: Charles H. Kerr, 1904), p. 33.

100. Karl Marx, *Critique of the Gotha Programme* (Moscow: Foreign Languages Publishing House, 1954), p. 16.

101. John B. Cobb, Jr., "The Population Explosion and the Rights of the Sub-human World," in *Environment and Society,* ed. Robert T. Roelofs, Joseph N. Crowley, and Donald L. Hardesty (Englewood Cliffs, N.J.: Prentice-Hall, 1974), p. 335.

102. *Capital,* vol. 1, p. 329. The idea recurs throughout Marx's writings.

103. Ibid., p. 390.

104. Ibid., p. 729.

105. Ibid., p. 730.

106. A. Leslie Banks and J. A. Hislop, "Sanitation Practices and Disease Control in Extending and Improving Areas for Human Habitation," in Thomas, *Man's Role in Changing the Face of the Earth,* vol. 2, p. 818.

107. Ernst Bloch, *Das Prinzip Hoffnung,* vol. 2 (Berlin: Aufbau Verlag, 1955), p. 246.

108. Alfred Schmidt, *The Concept of Nature in Marx,* trans. Ben Fowkes (London: NLB, 1971), p. 159.

109. Ibid., p. 160.

110. Ibid., p. 161.

111. Ibid., p. 162.

112. Ibid., p. 191.

113. Marx, *The Economic and Philosophic Manuscripts of 1844,* pp. 101-114.

114. Ibid., p. 102.

115. Lewis Mumford, *The Condition of Man* (New York: Harcourt, Brace, 1944), p. 339.

116. Ibid., p. 332.

117. *Reminiscences of Marx and Engels* (Moscow: Foreign Languages Publishing House, n. d.).

118. Marx to J. Weydemeyer, March 5, 1852, in *Selected Correspondence,* p. 86.

119. *Capital,* vol. 1, p. 334.

120. Ibid., p. 329.

121. Ibid., p. 326.

122. Ibid., pp. 326, 329.

123. Ibid., p. 333.

124. Ibid.

125. Ibid., p. 334.

126. Ibid., pp. 334-335.

127. *Capital,* vol. 3, p. 266.

128. Engels, *Dialectics of Nature,* pp. 402-405.

129. Ibid., p. 35.

130. Engels, *Dialectics of Nature,* p. 49.

131. Ibid., p. 406.

132. Ibid., pp. 234-235.

133. Cf. Adolf Portmann, "The Seeing Eye," in *The Subversive Science,* pp. 115-122.

134. Ibid. and Julian Huxley, *Evolution in Action* (New York: New American Library of World Literature, 1957), p. 29.

135. Letter to Arnold Ruge, May 1843; *A Contribution to the Critique of Political Economy*, p. 12; *Capital*, vol. 1, pp. 10, 751; *The Civil War in France* (Moscow: Foreign Languages Publishing House, 1952), p. 97; and *Critique of the Gotha Programme*, pp. 24, 26. See also Howard L. Parsons, *The Concept of Creativity in Marx* (Privately printed, 1976).

136. Marx, *The Economic and Philosophic Manuscripts of 1844*, p. 114.

137. Marx, *Grundrisse*, p. 202.

138. William Shakespeare, *Sonnets*, CXLVI.

139. Engels, *Dialectics of Nature*, p. 54.

140. René Dubos, *So Human An Animal* (New York: Charles Scribner's Sons, 1968), p. 107.

141. Fred Schwarz, *You Can Trust the Communists (to Be Communists)* Englewood Cliffs, N.J.: Prentice-Hall, 1962), p. 16.

142. Dirk Struik, Introduction to Marx, *The Economic and Philosophic Manuscripts of 1844*, pp. 47-56.

143. Jean-Paul Sartre, *The Problem of Method*, trans. Hazel E. Barnes (London: Methuen, 1963).

144. Robert Tucker, *Philosophy and Myth in Karl Marx* (New York: Cambridge University Press, 1964), pp. 239-240.

145. Louis Dupré, *The Philosophical Foundations of Marxism* (New York: Harcourt, Brace and World, 1966), p. 222.

146. Ibid., pp. 220-221.

147. Ibid., pp. 223-228.

148. Ibid., p. 230.

149. Ibid.

150. Marx and Engels, *Manifesto of the Communist Party*, p. 27.

151. Marx spoke of the "capitalized blood of children" (*Capital*, vol. 1, p. 756) and wrote that while "savages and animals have at least the need for hunting, exercise and companionship," children are turned into workers and the grown worker "has become a neglected child." Marx, *The Economic and Philosophic Manuscripts of 1844*, p. 149.

152. Marx, *A Contribution to the Critique of Political Economy*, p. 312.

153. For a contrast between the child-rearing practices in the United States and those in the Soviet Union, see Urie Bronfenbrenner, *Two Worlds of Childhood* (Washington, D.C.: Russell Sage Foundation, 1970).

154. Marx, Preparatory Notes to the Doctoral Dissertation. Karl Marx/Friedrich Engels, *Historisch critisch-Gesamtausgabe*, ed. D. Ryazanov, vol. 1, pt. 1 (Berlin: Marx-Engels Verlag, 1927-), p. 93.

155. *Capital*, vol. 1, p. 79.

156. John Bowlby, *Attachment and Loss*, vol. 1, *Attachment;* vol. 2, *Anxiety and Anger* (New York: Basic Books, 1969, 1973).

157. Francis Bacon, *Novum Organum*, Aphorisms, Books I, III. Cf. Book IV: "Toward the effecting of works, all that man can do is to put together or put asunder natural bodies. The rest is done by nature working within." Molecules act and interact in ways not fully understood or controlled by men; and in an analogous

and perhaps closely related way, it is possible to make automatic machines that in some respects "escape" the control of their human makers (Norbert Wiener). But these regions of man's uncontrol need not generate science fiction fears. While man's new-won control always creates new problems (if he creates living matter or prevents death, think of the problems), a wise control will keep in mind its purposes and its limits. For a discussion of the "moral and religious" implications of genetic control, cloning, artificial donor insemination, and related issues, see Paul Ramsey, *Fabricated Man: The Ethics of Genetic Control* (New Haven, Conn.: Yale University Press, 1970).

158. See John Somerville, "Ethics, Determinism, and Freedom," *Philosophy and Phenomenological Research* 28, no. 1 (September 1967): 20.

159. American Friends Service Committee, *Weapons for Counterinsurgency* (Philadelphia: NARMIC, 1970).

160. See Lewis Mumford, *The Myth of the Machine,* vol. 2, *The Pentagon of Power* (New York: Harcourt, Brace, Jovanivich, 1970), Plate 14.

161. Hugh H. Bennett, *Soil Conservation* (New York: McGraw-Hill Book Co., 1939).

162. H. H. Wooten, *Major Uses of Land in the United States,* Washington, D.C.: U.S. Department of Agriculture Technical Bulletin, No. 1082, 1953.

163. F. Fraser Darling and John P. Milton, eds., *Future Environments of North America* (Garden City, N.Y.: Doubleday, Natural History Press, 1966), p. 657.

164. Paul W. Richards, "The Tropical Rain Forest," *Scientific American* 229, no. 6 (December 1973): 58-67.

165. Barry Weisberg, *Ecocide in Indochina* (New York: Harper and Row, 1970).

166. M. King Hubbert, "The Energy Resources of the Earth," *Scientific American* 224, no. 3 (September 1971): 60-70.

167. Lester R. Brown, "Human Food Production as a Process in the Biosphere," *Scientific American* 223, no. 3 (September 1970): 161-170.

168. S. Fred Singer, "Human Energy Production as a Process in the Biosphere," *Scientific American* 223, no. 3 (September 1970): 175-190.

169. Brown, op. cit., pp. 161-170.

170. Commoner, op. cit., p. 13.

171. Handler, op. cit., p. 862. PCB, a toxic industrial chemical, is now widespread.

172. Gus Hall, op. cit., ch. 3. Hall cites *Hazards in the Industrial Environment,* the minutes of a 1969 conference called by the Oil, Chemical and Atomic Workers International Union.

173. Lawrence J. Henderson, *The Fitness of the Environment* (Boston: Beacon Press, 1958).

174. Theodor Rosebury, *Life on Man* (New York: Viking Press, 1969).

175. C. F. von Weizäcker, *The History of Nature,* trans. Fred D. Wieck (Chicago: University of Chicago Press, 1949), ch. 4.

176. Herrick, op. cit., ch. 5.

177. F. Fraser Darling, *Natural History in the Highlands and Islands* (London: Collins, 1955).

178. David E. Lilienthal, *TVA—Democracy on the March* (New York: Harper, 1944).

179. For the significance of "negative entropy" among organisms, see Ludwig

von Bertalanffy, *Robots, Men and Minds* (New York: George Braziller, 1967).

180. Handler, op. cit., p. 451.

181. Howard L. Parsons, *Man Today—Problems, Values, and Fulfillment*, vols. 4-5, *Revolutionary World* (Amsterdam: B. R. Grüner, 1974), ch. 8.

182. John E. Pfeiffer, *The Emergence of Man* (New York: Harper and Row, 1969), p. 10.

183. *Scientific American* 224, no. 3 (March 1971): 46-47.

184. Pfeiffer, op. cit., p. 99.

185. V. Gordon Childe, *Man Makes Himself* (New York: New American Library, 1951), p. 180.

186. Handler, op. cit., p. 902. China experienced a parallel population growth during the Manchu dynasty after 1650. Ta Chen, *Population in Modern China* (Chicago: University of Chicago Press, 1946).

187. Lewis Mumford, *Technics and Civilization* (New York: Harcourt, Brace, 1934).

188. C. P. Snow, *The Two Cultures and the Scientific Revolution* (New York: Cambridge University Press, 1961), pp. 30-31.

189. A. N. Whitehead, *Process and Reality* (New York: Macmillan Co., 1929), p. viii.

190. Harrison Brown, "Human Materials Production as a Process in the Biosphere" *Scientific American* 223, no. 3 (September 1970): 206.

191. Michael Harwood, "We Are Killing the Sea Around Us," *The New York Times Magazine*, October 24, 1971, p. 35.

192. Karl Marx and Frederick Engels, *Selected Correspondence 1846-1895*, trans. Dona Torr (New York: International Publishers, 1942), p. 204.

193. Ibid., pp. 235-237.

194. Nikolai Timofeyev-Ressovsky, "The Biosphere Is 10 Times Richer Than We Think," *The UNESCO Courier*, January 1973, pp. 29-31.

195. E. K. Fedorov and I. B. Novik, "Man and His Natural Environment," *Soviet Studies in Philosophy* 12, no. 2 (Fall 1973): 7-8. (This article originally appeared in *Voprosy filosofii*, no. 12, 1972.)

196. Ibid., p. 14.

197. Marx and Engels, *Selected Correspondence 1846-1895*, p. 199.

198. *Outlines of a Critique of Political Economy*, in Karl Marx, *The Economic and Philosophic Manuscripts of 1844*, p. 222.

199. Dr. Howard Reiquam, as reported in *The New York Times*, January 2, 1972.

200. Clarence J. Glacken, "Changing Ideas of the Habitable World," op. cit., p. 7.

201. Karl H. Niebyl, unpublished paper presented in the panel, "Some Theoretical Aspects of US-GDR Economic Relations," at the National Conference of the American Association for the Study of the German Democratic Republic, New York City, June 3, 1972.

202. *Imperialism: The Highest Stage of Capitalism* (New York: International Publishers, 1939).

203. *Capital*, Vol. 3, p. 678.

204. Marx, *A Contribution to the Critique of Political Economy*, p. 11.

205. J. D. Bernal, *The Social Function of Science* (Cambridge, Mass.: M.I.T.

Press, 1967), p. 28.

206. Handler, op. cit., chs. 15-19.

207. Ibid., pp. 894-896. The review by Waddington appeared in *The New York Times Book Review*, June 21, 1970, p. 2.

208. Ian L. McHarg, *Design with Nature* (Garden City, N.Y.: Doubleday, Natural History Press, 1969).

209. Loren Eiseley, "The Ethics of the Group," in *Social Control in a Free Society*, ed. Robert E. Spiller (Philadelphia: University of Pennsylvania Press, 1960), p. 30.

210. Pfeiffer, op. cit., p. 446.

211. *Capital*, vol. 1, p. 334.

212. *Capital*, vol. 3, p. 259.

213. *Capital*, vol. 1, p. 763.

214. Ibid., p. 335.

215. Ibid., p. 486.

216. V. I. Lenin, "On the Significance of Militant Materialism," in *Collected Works*, vol. 33 (Moscow: Progress Publishers, 1973), p. 229.

217. Contemporary research on innate human responses to forms, gestures, and movements in others are yielding evidence of our long sociality. The human organism is also coded to respond to diurnal rhythms, the changes of the seasons, lunar cycles, and probably other natural rhythms. Studies in the behavior of animals and then of man in their sensitivity to changes in nature are revealing how closely we are tied to the ecological patterns of the cosmos. See Dubos, op. cit., pp. 73-74. Dubos discusses man's evolutionary heritage from learned adaptations recorded in genetic structures and anatomy. So do Hugh H. Iltis, Orie L. Loucks, and Peter Andrews in "Criteria for an Optimum Human Environment," *Bulletin of the Atomic Scientists* 26, no. 1 (January 1970): 2-6.

218. Adolf Portmann, "The Seeing Eye"; Edith Cobb, "The Ecology of Imagination in Childhood"; and Grady Clay, "Remembered Landscapes," in *The Subversive Science*, pp. 115-139. Maxwell Weismann took ninety chronically hospitalized mental patients camping for two weeks, and within three months thirty-five returned to their communities. (Iltis, et al., op. cit.) For decades, Soviet scientists have stressed the therapy of work and physical exercise in the open. Cuba has had remarkable success in restoring health to once mentally ill persons by giving them opportunities for work of various kinds. The errors that urban administrative workers are prone to are no doubt reduced by having them cut cane, or, as in China, sending them into the countryside. See Mark G. Field, *Soviet Socialized Medicine* (New York: Free Press, 1967); Margaret Gilpin, "Health Care: Cuba," in *Cuba: People—Questions* (New York: Friendship, 1975), pp. 10-16; and Joshua S. Horn, *Away With All Pests: An English Surgeon in People's China, 1954-1969* (New York: Monthly Review, 1971).

219. *Capital*, vol. 1, p. 486.

220. Marx, *Grundrisse*, p. 119.

221. *Capital*, vol. 3, p. 250.

222. Ibid., p. 88.

223. Charles A. Scarlott, "Limitations to Energy Use," in Thomas, *Man's Role*

in Changing the Face of the Earth, vol. 2, pp. 1010-1022.

224. V. I. Lenin, *On Peaceful Coexistence* (Moscow: Progress Publishers, 1967).

225. Ibid., p. 27.

226. "Some 30 USA-USSR groups are already at work exchanging data on what can be done in both countries to save their lakes and rivers from pollution, to protect vast forest reserves from destruction, and fish, animal and bird life from extinction or serious depletion." Jessica Smith, "US-USSR Relations: A New Era Opens," *New World Review* 41, no. 3 (Summer 1973): 11. Such exchange among national groups should be greatly widened. See *The New York Times,* September 22, 1972.

227. In May 1958, Linus C. Pauling estimated that carbon-14 from bomb tests to that date would produce about one million seriously defective children and about two million embryonic and neonatal deaths (*The New York Times,* August 11, 1958, p. 60). Soviet scientists made corresponding estimates of Soviet explosions. The report of the United Nations Scientific Committee on the Effects of Atomic Radiation stressed the impairment of man's "genetic inheritance" by radiation and the responsibility for correcting it—though the fifteen-nation committee rejected the Soviet proposal for "immediate cessation" of nuclear tests (*The New York Times,* August 11, 1958, p. 6).

228. John Somerville, *The Peace Revolution: Ethos and Social Process* (Westport, Conn.: Greenwood Press, 1975).

229. William M. Mandel, in an effort to correct false statements in *The New York Times,* has cited evidence for a mass-based ecological movement in the U.S.S.R.: a successful nineteen-month public struggle against the building of a power dam and a huge reservoir; forest conservation; reduction in acreage of land condemned for industrial use; decrease of discharge of untreated effluents into bodies of water; a growing inland fish catch and phenomenal increase in wildlife; prevention of dust storms of the magnitude of the one in 1928, in a land of very variable precipitation and high hot winds; the absence of photochemical fog and of the use of leaded gasoline in the cities; and the popular campaign to block the commercial pollution of Lake Baikal. See his "Environment, USSR," *New World Review* 41, no. 1 (1st quarter, 1973): 78-80. A longer article by Mandel, with a copious bibliography pro and con, Soviet and American, is "The Soviet Ecology Movement," *Science and Society* 36, no. 4 (Winter 1972): 385-416.

230. Ivan Yefremov, "Are We Alone in Outer Space?" *Soviet Life* 151 (April 1969): 26.

231. Report to Joint Jubilee Meeting of the CPSU, Central Committee, Supreme Soviet of the U.S.S.R. and Supreme Soviet of the Russian Federation, Moscow, November 3, 1967, in *October Revolution 1917-1967* (Moscow: Novosti, n. d.), p. 49.

232. Leo A. Orleans and Richard P. Suttmeier, "The Mao Ethics of Environmental Quality," *Science* 170 (December 11, 1970): 1173-1176.

233. After a visit to the U.S.S.R., John A. Creedy, president of the U.S. Water Transportation Association, said: "The most impressive thing to me was that the Russians appear to have licked the pollution problem." Quoted in *New World Review* 41, no. 3 (Summer 1973): 11, from *The New York Times,* July 8, 1973. A 1974 Sweden-Norway-Finland-Denmark convention permits citizens' environmental suits in any other country.

part 2
Selections from the Writings of Marx and Engels on Ecology

I. The Matter of Nature as Prior and Prerequisite to Man's Labor

Contrary to the bourgeoisie and those unduly influenced by the bad press given through the bourgeoisie to the ideas of Marx and Engels, the latter never denied the independent power of nonhuman nature. Marx and Engels repeatedly pointed to the "objective condition" of man's natural and historical environment within which he lives, labors, and has his being. Nature is an existence which Marx aptly described as "presupposed" for man's communal activity. Through its instruments of labor, materials, climate, and other characteristics, nature determines man's production and in turn is determined by that production. —Ed.

A. Man Does Not Create Matter

"The system of appropriation through work," Proudhon goes on, "is therefore *contrary to law*; and when the supporters of that system claim it as an explanation of their laws they are *contradicting themselves.*"

To say further, according to this opinion, that the cultivation of the land, for example, "creates fullest ownership of the same" is *a petitio principii.* It is a fact that a new productive capacity of matter has been created. But what was to be proved was that ownership of matter itself was thus created. Man has not created matter itself. And he cannot even create any productive capacity if the matter does not exist beforehand.

K. Marx and F. Engels, *The Holy Family,* p. 65.

B. *Matter Is Prerequisite to Labor*

It is wrong to speak of labor in so far as it is applied to the production of use-values as of the only source of wealth, namely, the material wealth produced by it. Being an activity intended to adapt materials to this or that purpose, it requires matter as a prerequisite. In different use-values the proportion between labor and raw material varies greatly, but use-value always has a natural substratum. Labor, as an activity, directed to the adaptation of raw material in one form or another, is a natural condition of human existence, a condition of exchange of matter between man and nature, independent of all social forms.

Karl Marx, *A Contribution to the Critique of Political Economy,* p. 33.

C. *Matter Independent of Labor Has No Value*

The purely natural material in which *no* human labour is objectified, to the extent that it is merely a material that exists independently of labour, has no value; as little value as is possessed by the common elements as such.

Karl Marx, *Grundrisse,* p. 366.

D. *The Earth Is the Objective and Presupposed Condition of Man's Reproduction and Development Through Communal Labor*

The main point here is this: In all these forms—in which landed property and agriculture form the basis of the economic order, and where the economic aim is hence the production of use-values, i.e., the *reproduction of the individual* within the specific relation to the commune in which he is its basis—there is to be found: (1) Appropriation not through labour, but presupposed to labour; appropriation of the natural conditions of labour, of the *earth* as the original instrument of labour as well as its workshop and repository of raw materials. The individual relates simply to the objective conditions of labour as being his; [relates] to them as the inorganic nature of his subjectivity, in which the latter realizes itself; the chief objective condition of labour does not itself appear as a *product* of labour, but is already there as *nature*; on one side the living individual, on the other the earth, as the objective condition of his reproduction; (2) but this *relation* to land and soil, to the earth, as the property of the labouring individual—who thus appears from the outset not merely as labouring individual, in this abstraction, but who has an *objective mode of existence* in his ownership of the land, an existence *presupposed* to his activity, and not merely as a result of it, a presupposition of his activity just like his skin, his sense organs, which of course he also reproduces and develops etc. in the life process of his reproduction—is instantly mediated by the naturally arisen, spontaneous, more or less historically developed and modified presence of the individual as *member of a commune*—his naturally arisen presence as member of a tribe etc. An isolated individual could no more have property in land and soil than he could speak. He could, of course, live off it as substance, as do the animals. The relation to the earth as property is always mediated through the occupation of the land and soil, peacefully or violently, by the tribe, the commune, in some more or less naturally arisen or already historically developed form.

Karl Marx, Grundrisse, p. 485.

E. The Influence of Natural Conditions on the Productiveness of Labor

If the labourer wants all his time to produce the necessary means of subsistence for himself and his race, he has no time left in which to work gratis for others. Without a certain degree of productiveness in his labour, he has no such superfluous time at his disposal; without such superfluous time no surplus-labour, and therefore no capitalists, no slave-owners, no feudal lords, in one word, no class of large proprietors.[1]

Thus we may say that surplus-value rests on a natural basis; but this is permissible only in the very general sense, that there is no natural obstacle absolutely preventing one man from disburdening himself of the labour requisite for his own existence, and burdening another with it, any more, for instance, than unconquerable natural obstacles prevent one man from eating the flesh of another.[2] No mystical ideas must in any way be connected, as sometimes happens, with this historically developed productiveness of labour. It is only after men have raised themselves above the rank of animals, when therefore their labour has been to some extent socialised, that a state of things arises in which the surplus-labour of the one becomes a condition of existence for the other. At the dawn of civilisation the productiveness acquired by labour is small, but so too are the wants which develop with and by the means of satisfying them. Further, at that early period, the portion of society that lives on the labour of others is infinitely small compared with the mass of direct producers. Along with the progress in the productiveness of labour, that small portion of society increases both absolutely and relatively.[3] Besides, capital with its accompanying relations springs up from an economic soil that is the product of a long process of development. The productiveness of labour that serves as its foundation and starting-point, is a gift, not of Nature, but of a history embracing thousands of centuries.

Apart from the degree of development, greater or less, in the form of social production, the productiveness of labour is fettered by physical

---

1. "The very existence of the master-capitalists, as a distinct class, is dependent on the productiveness of industry." (Ramsay, 1. c., p. 206.) "If each man's labour were but enough to produce his own food, there could be no property." (Ravenstone, 1. c., pp. 14, 15.)

2. According to a recent calculation, there are yet at least 4,000,000 cannibals in those parts of the earth which have already been explored.

3. "Among the wild Indians in America, almost everything is the labourer's, 99 parts of a hundred are to be put upon the account of labour. In England, perhaps, the labourer has not 2/3." ("The Advantages of the East India Trade, &c.," p. 73.)

conditions. These are all referable to the constitution of man himself (race, &c.), and to surrounding Nature. The external physical conditions fall into two great economic classes, (1) Natural wealth in means of subsistence, i.e., a fruitful soil, waters teeming with fish, &c., and (2), natural wealth in the instruments of labour, such as waterfalls, navigable rivers, wood, metal, coal, &c. At the dawn of civilisation, it is the first class that turns the scale; at a higher stage of development, it is the second. Compare, for example, England with India, or in ancient times, Athens and Corinth with the shores of the Black Sea.

The fewer the number of natural wants imperatively calling for satisfaction, and the greater the natural fertility of the soil and the favourableness of the climate, so much less is the labour-time necessary for the maintenance and reproduction of the producer. So much greater therefore can be the excess of his labour for others over his labour for himself. Diodorus long ago remarked this in relation to the ancient Egyptians. "It is altogether incredible how little trouble and expense the bringing up of their children causes them. They cook for them the first simple food at hand; they also give them the lower part of the papyrus stem to eat, so far as it can be roasted in the fire, and the roots and stalks of marsh plants, some raw, some boiled and roasted. Most of the children go without shoes and unclothed, for the air is so mild. Hence a child, until he is grown up, costs his parents no more, on the whole, than twenty drachmas. It is this, chiefly, which explains why the population of Egypt is so numerous, and, therefore, why so many great works can be undertaken."[4] Nevertheless the grand structures of ancient Egypt are less due to the extent of its population than to the large proportion of it that was freely disposable. Just as the individual labourer can do more surplus-labour in proportion as his necessary labour-time is less, so with regard to the working population. The smaller the part of which is required for the production of the necessary means of subsistence, so much the greater is the part that can be set to do other work.

Capitalist production once assumed, then, all other circumstances remaining the same, and given the length of the working-day, the quantity of surplus-labour will vary with the physical conditions of labour, espe-

---

4. Diodorus, l. c., l. 1., c. 80.

cially with the fertility of the soil. But it by no means follows from this that the most fruitful soil is the most fitted for the growth of the capitalist mode of production. This mode is based on the dominion of man over Nature. Where Nature is too lavish, she "keeps him in hand, like a child in leading-strings." She does not impose upon him any necessity to develop himself.[5] It is not the tropics with their luxuriant vegetation, but the temperate zone, that is the mother-country of capital. It is not the mere fertility of the soil, but the differentiation of the soil, the variety of its natural products, the changes of the seasons, which form the physical basis for the social division of labour, and which, by changes in the natural surroundings, spur man on to the multiplication of his wants, his capabilities, his means and modes of labour. It is the necessity of bringing a natural force under the control of society, of economising, of appropriating or subduing it on a large scale by the work of man's hand, that first plays the decisive part in the history of industry. Examples are, the irrigation works in Egypt,[6] Lombardy, Holland, or in India and Persia where irrigation by means of artificial canals, not only supplies the soil with the water indispensable to it, but also carries down to it, in the shape of sediment from the hills, mineral fertilisers. The secret of the flourishing state

---

5. "The first (natural wealth) as it is most notable and advantageous, so doth it make the people careless, proud, and given to all excesses; whereas the second enforceth vigilancy, literature, arts and policy." ("England's Treasure by Foreign Trade. Or the Balance of our Foreign Trade is the Rule of our Treasure. Written by Thomas Mun of London, merchant, and now published for the common good by his son John Mun." London, 1969, pp. 181, 182.) "Nor can I conceive a greater curse upon a body of people, than to be thrown upon a spot of land, where the productions for subsistence and food were, in great measure, spontaneous, and the climate required or admitted little care for raiment and covering . . . there may be an extreme on the other side. A soil incapable of produce by labour is quite as bad as a soil that produces plentifully without any labour." ("An Enquiry into the Causes of the Present of High Price of Provisions." Lond. 1767, p. 10.)

6. The necessity for predicting the rise and fall of the Nile created Egyptian astronomy, and with it the dominion of the priests, as directors of agriculture. "Le solstice est le moment de l'année où commence la crue du Nil, et celui que les Egyptiens ont dû observer avec le plus d'attention. . . . C'était cette année tropique qu'il leur importait de marquer pour se diriger dans leurs operations agricoles. Ils durent donc chercher dans le ciel un signe apparent de son retour." (Cuvier: "Discours sur les révolutions du globe," ed. Hoefer, Paris, 1863, p. 141.)

of industry in Spain and Sicily under the dominion of the Arabs lay in their irrigation works.[7]

Favourable natural conditions alone, give us only the possibility, never the reality, of surplus-labour, nor, consequently, of surplus-value and a surplus-product. The result of difference in the natural conditions of labour is this, that the same quantity of labour satisfies, in different countries, a different mass of requirements,[8] consequently, that under circumstances in other respects analogous, the necessary labour-time is different. These conditions affect surplus-labour only as natural limits, *i.e.,* by fixing the points at which labour for others can begin. In proportion as industry advances, these natural limits recede. In the midst of our West European society, where the labourer purchases the right to work for his own livelihood only by paying for it in surplus-labour, the idea easily takes root that it is an inherent quality of human labour to furnish a surplus-product.[9] But consider, for example, an inhabitant of the eastern islands of the Asiatic Archipelago, where sago grows wild in the forests. "When the inhabitants have convinced themselves, by boring a hole in the tree, that the pith is ripe, the trunk is cut down and divided into several pieces, the pith is extracted, mixed with water and filtered: it is then quite fit

---

7. One of the material bases of the power of the State over the small disconnected producing organisms in India, was the regulation of the water supply. The Mahometan rulers of India understood this better than their English successors. It is enough to recall to mind the famine of 1866, which cost the lives of more than a million Hindus in the district of Orissa, in the Bengal presidency.

8. "There are no two countries which furnish an equal number of the necessaries of life in equal plenty, and with the same quantity of labour. Men's wants increase or diminish with the severity or temperateness of the climate they live in; consequently, the proportion of trade which the inhabitants of different countries are obliged to carry on through necessity cannot be the same, nor is it practicable to ascertain the degree of variation farther than by the degrees of Heat and Cold; from whence one may make this general conclusion, that the quantity of labour required for a certain number of people is greatest in cold climates, and least in hot ones; for in the former men not only want more clothes, but the earth more cultivating than in the latter." ("An Essay on the Governing Causes of the Natural Rate of Interest." Lond. 1750, p. 59.) The author of this epoch-making anonymous work is J. Massie. Hume took his theory of interest from it.

9. "Chaque travail doit (this appears also to be part of the droits et devoirs du citoyen) laisser un excédant." Proudhon.

for use as sago. One tree commonly yields 300 lbs., and occasionally 500 to 600 lbs. There, then, people go into the forests, and cut bread for themselves just as with us they cut fire-wood."[10] Suppose now such an eastern bread-cutter requires 12 working-hours a week for the satisfaction of all his wants. Nature's direct gift to him is plenty of leisure time. Before he can apply this leisure time productively for himself, a whole series of historical events is required; before he spends it in surplus-labour for strangers, compulsion is necessary. If capitalist production were introduced the honest fellow would perhaps have to work six days a week, in order to appropriate to himself the product of one working-day. The bounty of Nature does not explain why he would then have to work 6 days a week, or why he must furnish 5 days of surplus-labour. It explains only why his necessary labour-time would be limited to one day a week. But in no case would his surplus-product arise from some occult quality inherent in human labour.

Thus, not only does the historically developed social productiveness of labour, but also its natural productiveness, appear to be productiveness of the capital with which that labour is incorporated.

Karl Marx, *Capital*, vol. 1, pp. 511-515.

F. Why Landed Property Never Developed in the East: Climate and Soil

... The absence of property in land is indeed the key to the whole of the East. Herein lies its political and religious history. But how does it come about that the Orientals did not arrive at landed property, even in its feudal form? I think it is mainly due to the climate, taken in connection with the nature of the soil, especially with the great stretches of desert which extend from the Sahara straight across Arabia, Persia, India and Tatary up to the highest Asiatic plateau. Artificial irrigation is here the first condition of agriculture and this is a matter either for the communes, the provinces or the central government. An oriental government never had more than three departments: finance (plunder at home), war (plunder at home and abroad), and public works (provision for reproduction). The British Government in India has administered Nos. 1 and 2 in a rather

---

10. F. Schouw: "Die Erde, die Pflanzen und der Mensch," 2 Ed. Leipz. 1854, p. 148.

narrow-minded spirit and dropped No. 3 entirely, so that Indian agriculture is being ruined. Free competition discredits itself there completely. This artificial fertilization of the land, which immediately ceased when the irrigation system fell into decay, explains the otherwise curious fact that whole stretches which were once brilliantly cultivated are now waste and bare (Palmyra, Petra, the ruins in the Yemen, districts in Egypt, Persia and Hindustan): it explains the fact that one single devastating war could depopulate a country for centuries and strip it of its whole civilization.

> Letter of Engels to Marx, June 6, 1853,
> Karl Marx and Frederick Engels,
> *Selected Correspondence,* p. 99.

II. Nature as Dialectical

Man's collective and immediate material dealings with nature bring him into a dialectical relation with it, into a dynamic and potentially developing interaction with it. Such a relation, when critically analyzed, reveals nature as continuous motion, interconnection, and transformation. Nature is a ceaseless series of unities of opposites, which are mutually creative, mutually destructive, and mutually transforming.   —Ed.

Dialectics Versus Metaphysics

When we reflect on Nature, or the history of mankind, or our own intellectual activity, the first picture presented to us is of an endless maze of relations and interactions, in which nothing remains what, where and as it was, but everything moves, changes, comes into being and passes out of existence. This primitive, naive, yet intrinsically correct conception of the world was that of ancient Greek philosophy, and was first clearly formulated by Heraclitus: everything is and also is not, for everything is in *flux,* is constantly changing, constantly coming into being and passing away. But this conception, correctly as it covers the general character of the picture of phenomena as a whole, is yet inadequate to explain the details of which this total picture is composed; and so long as we do not understand these, we also have no clear idea of the picture as a whole. In order to understand these details, we must detach them from their natural or historical connections, and examine each one separately, as to its nature, its special causes and effects, etc. This is primarily the task of natural science and historical research; branches of science which the Greeks of the classical period, on very good grounds, relegated to a merely subordinate position, because they had first of all to collect materials for these to work upon. The beginnings of the exact investigation of nature were first developed by the Greeks of the Alexandrian period, and later on, in the Middle Ages, were further developed by the Arabs. Real natural science, however, dates only from the second half of the fifteenth century, and from then on it has advanced with constantly increasing rapidity.

The analysis of Nature into its individual parts, the grouping of the different natural processes and natural objects in definite classes, the study of the internal anatomy of organic bodies in their manifold forms—these were the fundamental conditions of the gigantic strides in our knowledge of Nature which have been made during the last four hundred years. But this method of investigation has also left us as a legacy the habit of observing natural objects and natural processes in their isolation, detached from the whole vast interconnection of things; and therefore not in their motion, but in their repose; not as essentially changing, but as fixed constants; not in their life, but in their death. And when, as was the case with Bacon and Locke, this way of looking at things was transferred from natural science to philosophy, it produced the specific narrow-mindedness of the last centuries, the metaphysical mode of thought.

To the metaphysician, things and their mental images, ideas, are isolated, to be considered one after the other apart from each other, rigid, fixed objects of investigation given once for all. He thinks in absolutely discontinuous antitheses. His communication is: "Yea, yea, Nay, nay, for whatsoever is more than these cometh of evil." For him a thing either exists, or it does not exist; it is equally impossible for a thing to be itself and at the same time something else. Positive and negative absolutely exclude one another; cause and effect stand in an equally rigid antithesis one to the other. At first sight this mode of thought seems to us extremely plausible, because it is the mode of thought of so-called sound common sense. But sound common sense, respectable fellow as he is within the homely precincts of his own four walls, has most wonderful adventures as soon as he ventures out into the wide world of scientific research. Here the metaphysical mode of outlook, justifiable and even necessary as it is in domains whose extent varies according to the nature of the object under investigation, nevertheless sooner or later always reaches a limit beyond which it becomes one-sided, limited, abstract, and loses its way in insoluble contradictions. And this is so because in considering individual things it loses sight of their connections; in contemplating their existence it forgets their coming into being and passing away; in looking at them at rest it leaves their motion out of account; because it cannot see the wood for the trees. For everyday purposes we know, for example, and can say with certainty whether an animal is alive or not; but when we look more closely we find that this is often an extremely complex question, as jurists know very well. They have cudgelled their brains in vain to discover some rational limit beyond which the killing of a child in its mother's womb is murder; and it is equally impossible to determine the moment of death, as physiology has established that death is not a sudden, instantaneous event, but a very protracted process. In the same way every organic being is at each moment the same and not the same; at each moment it is assimilating matter drawn from without, and excreting other matter; at each moment the cells of its body are dying and new ones are being formed; in fact, within a longer or shorter period the matter of its body is completely renewed and is replaced by other atoms of matter, so that every organic being is at all times itself and yet something other than itself. Closer investigation also shows us that the two poles of an antithesis, like positive and negative, are just as inseparable from each other as they are opposed, and that despite all their opposition they mutually penetrate each other. It is just the same with cause and effect; these are conceptions which only have validity in their application to a particular case as such, but when we consider the particular case in its general connection with the world as a whole they merge and dissolve in the conception of universal action and interaction, in which causes and effects are constantly changing places, and what is now or here an effect becomes there or then a cause, and *vice versa.*

None of these processes and methods of thought fit into the frame of metaphysical thinking. But for dialectics, which grasps things and their images, ideas, essentially in their interconnection, in their sequence, their movement, their birth and death, such processes as those mentioned above are so many corroborations of its own method of treatment. Nature is the test of dialectics, and it must be said for modern natural science that it has furnished extremely rich and daily increasing materials for this test, and has thus proved that in the last analysis Nature's process is dialectical and not metaphysical. But the scientists who have learnt to think dialectically are still few and far between, and hence the conflict between the discoveries made and the old traditional mode of thought is the explanation of the boundless confusion which now reigns in theoretical natural science and reduces both teachers and students, writers and readers to despair.

An exact representation of the universe, of its evolution and that of mankind, as well as of the reflection of this evolution in the human mind, can therefore only be built up in a dialectical way, taking constantly into account the general actions and reactions of becoming and ceasing to be, of progressive or retrogressive changes.

Frederick Engels, Herr Eugen Dühring's
Revolution in Science, pp. 26-29.

III. The Interdependence of Man as a Living Being with Nature

Man depends on nature for vital substances (such as food, water, and oxygen) and processes (such as photosynthesis), and nature in turn is affected by man's activities, such as the use of fire and the domestication of plants and animals. Marx described man's interdependence with nature at both the biochemical and the psychological levels. —Ed.

A. Nature Is Man's Body, on Which He Lives

The life of the species, both in man and in animals, consists physically in the fact that man (like the animal) lives on inorganic nature; and the more universal man is compared with an animal, the more universal is the sphere of inorganic nature on which he lives. Just as plants, animals, stones, air, light, etc., constitute theoretically a part of human consciousness, partly as objects of natural science, partly as objects of art—his spiritual inorganic nature, spiritual nourishment which he must first prepare to make palatable and digestible—so also in the realm of practice they constitute a part of human life and human activity. Physically man lives only on these products of nature, whether they appear in the form of food, heating, clothes, a dwelling, etc. The universality of man appears in practice precisely in the universality which makes all nature his *inorganic* body—both inasmuch as nature is (1) his direct means of life, and (2) the material, the object, and the instrument of his life activity. Nature is man's *inorganic body*—nature, that is, in so far as it is not itself the human body. Man *lives* on nature—means that nature is his *body*, with which he must remain in continuous interchange if he is not to die. That man's physical and spiritual life is linked to nature means simply that nature is linked to itself, for man is a part of nature.

> Karl Marx, *The Economic and Philo-*
> *sophic Manuscripts of 1844*, p. 112.

B. Man's Essential Interdependence with Nature

Whenever real, corporeal *man,* man with his feet firmly on the solid ground, man exhaling and inhaling all the forces of nature, *establishes* his real, objective *essential powers* as alien objects by his externalization, it is not the *act of positing* which is the subject in this process: it is the subjectivity of *objective* essential powers, whose action, therefore, must also be something *objective.* An objective being acts objectively, and he would not act objectively if the objective did not reside in the very nature of his being. He creates or establishes only *objects, because* he is established by objects—because at bottom he is *nature.* In the act of establishing, therefore, this objective being does not fall from his state of "pure activity" into a *creating of the object*; on the contrary, his *objective* product only confirms his *objective* activity, establishing his activity as the activity of an objective, natural being.

Here we see how consistent naturalism or humanism distinguishes itself both from idealism and materialism, constituting at the same time the unifying truth of both. We see also how only naturalism is capable of comprehending the act of world history.

Man is directly a *natural being.* As a natural being and as a living natural being he is on the one hand endowed with *natural powers of life*—he is an *active* natural being. These forces exist in him as tendencies and abilities—

as *instincts.* On the other hand, as a natural corporeal, sensuous, objective being he is a *suffering,* conditioned and limited creature, like animals and plants. That is to say, the *objects* of his instincts exist outside him, as *objects* independent of him; yet these objects are *objects* that he *needs*—essential *objects,* indispensable to the manifestation and confirmation of his essential powers. To say that man is a *corporeal,* living, real, sensuous, objective being full of natural vigor is to say that he has *real, sensuous, objects* as the objects of his life, or that he can only express his life in real, sensuous objects. *To be* objective, natural and sensuous, and at the same time to have object, nature and sense outside oneself, or oneself to be object, nature and sense for a third party, is one and the same thing. *Hunger* is a natural *need;* it therefore needs a *nature* outside itself, an *object* outside itself, in order to satisfy itself, to be stilled. Hunger is an acknowledged need of my body for an *object* existing outside it, indispensable to its integration and to the expression of its essential being. The sun is the *object* of the plant—an indispensable object to it, confirming its life—just as the plant is an object of the sun, being an *expression* of the life-awakening power of the sun, of the sun's *objective* essential power.

A being which does not have its nature outside itself is not a *natural* being, and plays no part in the system of nature. A being which has no object outside itself is not an objective being. A being which is not itself an object for some third being has no being for its *object;* i.e., it is not objectively related. Its be-ing is not objective.

An unobjective being is a *nullity*—an *un-being.*

Suppose a being which is neither an object itself, nor has an object. Such a being, in the first place, would be the *unique* being: there would exist no being outside it—it would exist solitary and alone. For as soon as there are objects outside me, as soon as I am not *alone,* I am *another—another reality* than the object outside me. For this third object I am thus an *other reality* than itself; that is, I am *its* object. Thus, to suppose a being which is not the object of another being is to presuppose that *no* objective being exists. As soon as I have an object, this object has me for an object. But a *non-objective* being is an unreal, nonsensical thing—a product of mere thought (hence of mere imagination)—a creature of abstraction. To be *sensuous* is to be an object of sense, to be a *sensuous* object, and thus to have sensuous objects outside oneself—objects of one's sensuousness. To be sensuous is to *suffer.*

Man as an objective, sensuous being is therefore a *suffering* being—and because he feels what he suffers, a *passionate* being. Passion is the essential force of man energetically bent on its object.

But man is not merely a natural being: he is a *human* natural being. That is to say, he is a being for himself. Therefore he is a *species being,* and has to confirm and manifest himself as such both in his being and in his knowing. Therefore, *human* objects are not natural objects as they immediately present themselves, and neither is *human sense* as it immediately *is*—as it is objectively—*human* sensibility, human objectivity. Neither nature objectively nor nature subjectively is directly given in a form adequate to the *human* being. And as everything natural has to have its *beginning, man* too has his act of origin—*history*—which, however, is for him a known history, and hence as an act of origin it is a conscious self-transcending act of origin. History is the true natural history of man (on which more later).

> Karl Marx, *The Economic and Philo-sophic Manuscripts of 1844,* pp. 180-182.

C. *The Mutual Affirmation of Man and His Objects*

If man's *feelings,* passions, etc., are not merely anthropological phe-nomena in the [narrower] sense, but truly *ontological* affirmations of being (of nature), and if they are only really affirmed because their *ob-ject* exists for them as a sensual object, then it is clear:

(1) That they have by no means merely one mode of affirmation, but rather that the distinct character of their existence, of their life, is con-stituted by the distinct mode of their affirmation. In what manner the object exists for them, is the characteristic mode of their *gratification.*

(2) Wherever the sensuous affirmation is the direct annulment of the object in its independent form (as in eating, drinking, working up of the object, etc.), this is the affirmation of the object.

(3) In so far as man, and hence also his feeling, etc., are *human,* the affirmation of the object by another is likewise his own gratification.

(4) Only through developed industry—i.e., through the medium of private property—does the ontological essence of human passion come into being, in its totality as in its humanity; the science of man is there-fore itself a product of man's establishment of himself by practical activity.

(5) The meaning of private property—apart from its estrangement—

is the *existence of essential objects* for man, both as objects of gratification and as objects of activity.

> Karl Marx, *The Economic and Philosophic Manuscripts of 1844*, p. 165.

D. Matter as Motion, Creativity, and Sensuous Quality

The real founder of *English materialism* and all *modern experimental* science was *Bacon.* For him natural science was true science and *physics* based on perception was the most excellent part of natural science. *Anaxagoras* with his *homoeomeria* and *Democritus* with his atoms are often the authorities he refers to. According to his teaching the *senses* are infallible and are the *source* of all knowledge. Science is *experimental* and consists in applying a *rational method* to the data provided by the senses. Induction, analysis, comparison, observation and experiment are the principal requisites of rational method. The first and most important of the inherent qualities of *matter* is *motion,* not only *mechanical* and *mathematical* movement, but still more *impulse, vital life-spirit, tension,* or, to use Jacob Bohme's expression, the throes [*Qual*] of matter. The primary forms of matter are the living, individualizing *forces of being* inherent in it and producing the distinction between the species.

In *Bacon,* its first creator, materialism contained latent and still in a naive way the germs of all-round development. Matter smiled at man with poetical sensuous brightness. The aphoristic doctrine itself, on the other hand, was full of the inconsistencies of theology.

> K. Marx and F. Engels, *The Holy Family,* p. 172.

IV. Man's Interdependence with Nature as a Being That Makes a Living

As distinct from other living beings, man not only "lives on inorganic nature," but he also makes a living of nature, interacting with it by means of his brain, hand, and tools in order to subsist. Thus, far more than any other species, the human species puts its "stamp on nature." Marx and Engels addressed themselves to the question of how the human species is

differentiated from other animals, because on the one hand they wished
to refute the idealists who repudiated man's animal nature and because
on the other hand they wanted to demonstrate the great possibilities for
fulfillment in man's evolutionary relation to nature.   —Ed.

A. By Productively Interacting with Nature, Man Subsists

The first premise of all human history is, of course, the existence of
living human individuals. Thus the first fact to be established is the physical
organization of these individuals and their consequent relation to the rest
of nature. Of course, we cannot here go either into the actual physical
nature of man, or into the natural conditions in which man finds himself—
geological, oro-hydrographical, climatic and so on. The writing of history
must always set out from these natural bases and their modification in
the course of history through the action of man.

Men can be distinguished from animals by consciousness, by religion
or anything else you like. They themselves begin to distinguish themselves
from animals as soon as they begin to *produce* their means of subsistence,
a step which is conditioned by their physical organization. By producing
their means of subsistence men are indirectly producing their actual
material life.

The way in which men produce their means of subsistence depends
first of all on the nature of the actual means they find in existence and
have to reproduce. This mode of production must not be considered
simply as being the reproduction of the physical existence of the
individuals. Rather it is a definite form of activity of these individuals,
a definite form of expressing their life, a definite *mode of life* on their
part. As individuals express their life, so they are. What they are, therefore, coincides with their production, both with *what* they produce and
with *how* they produce. The nature of individuals thus depends on the
material conditions determining their production.

Karl Marx and Frederick Engels,
The German Ideology, p. 7.

B. Man Has Impressed His Stamp on Nature Through Hand and Brain

When after thousands of years of struggle the differentiation of hand
from foot, and erect gait, were finally established, man became distinct

from the ape and the basis was laid for the development of articulate
speech and the mighty development of the brain that has since made the
gulf between man and the ape an unbridgeable one. The specialization
of the hand—this implies the *tool,* and the tool implies specific human
activity, the transforming reaction of man on nature, production. Animals
in the narrower sense also have tools, but only as limbs of their bodies:
the ant, the bee, the beaver; animals also produce, but their productive
effect on surrounding nature in relation to the latter amounts to nothing
at all. Man alone has succeeded in impressing his stamp on nature, not only
by so altering the aspect and climate of his dwelling-place, and even the
plants and animals themselves, that the consequences of his activity can
disappear only with the general extinction of the terrestrial globe. And
he has accomplished this primarily and essentially by means of *the hand.*
Even the steam-engine, so far his most powerful tool for the transformation of nature, depends, because it is a tool, in the last resort on the hand.
But step by step with the development of the hand went that of the brain;
first of all came consciousness of the conditions for separate practically
useful actions, and later, among the more favoured peoples and arising
from that consciousness, insight into the natural laws governing them.
And with the rapidly growing knowledge of the laws of nature the means
for reacting on nature also grew; the hand alone would never have achieved
the steam-engine if, along with and parallel to the hand, and partly owing
to it, the brain of man had not correspondingly developed.

   With man we enter *history.* Animals also have a history, that of their
descent and gradual evolution to their present position. This history, however, is made for them, and in so far as they themselves take part in it,
this occurs without their knowledge and desire. On the other hand, the
more that human beings become removed from animals in the narrower
sense of the word, the more they make their history themselves, consciously,
the less becomes the influence of unforeseen effects and uncontrolled
forces on this history, and the more accurately does the historical result
correspond to the aim laid down in advance.

                                   Frederick Engels, *Dialectics of Nature,*
                                   pp. 46-48.

C. Man's Reaction on Nature

   Natural science, like philosophy, has hitherto entirely neglected the
influence of men's activity on their thought; both know only nature on

the one hand and thought on the other. But it is precisely *the alteration of nature by men,* not solely nature as such, which is the most essential and immediate basis of human thought, and it is in the measure that man has learned to change nature that his intelligence has increased. The naturalistic conception of history, as found, for instance, to a greater or lesser extent in Draper and other scientists, as if nature exclusively reacts on man, and natural conditions everywhere exclusively determined his historical development, is therefore one-sided and forgets that man also reacts on nature, changing it and creating new conditions of existence for himself. There is devilishly little left of "nature" as it was in Germany at the time when the Germanic peoples immigrated into it. The earth's surface, climate, vegetation, fauna, and the human beings themselves have infinitely changed, and all this owing to human activity, while the changes of nature in Germany which have occurred in this period of time without human interference are incalculably small.

Frederick Engels, Dialectics of Nature,
p. 306.

D. The Laws of Human Life Are Different from the Laws of Animal Life

The struggle for life. Until Darwin, what was stressed by his present adherents was precisely the harmonious co-operative working of organic nature, how the plant kingdom supplies animals with nourishment and oxygen, and animals supply plants with manure, ammonia, and carbonic acid. Hardly was Darwin recognized before these same people saw everywhere nothing but *struggle.* Both views are justified within narrow limits, but both are equally one-sided and prejudiced. The interaction of bodies in non-living nature includes both harmony and collisions, that of living bodies conscious and unconscious co-operation as well as conscious and unconscious struggle. Hence, even in regard to nature, it is not permissible one-sidedly to inscribe only "struggle" on one's banners. But it is absolutely childish to desire to sum up the whole manifold wealth of historical evolution and complexity in the meagre and one-sided phrase "struggle for existence." That says less than nothing.

The whole Darwinian theory of the struggle for existence is simply the transference from society to organic nature of Hobbes' theory of *bellum omnium contra omnes* and of the bourgeois economic theory of competition, as well as the Malthusian theory of population. When once this feat has been accomplished (the unconditional justification for which, especially as regards the Malthusian theory, is still very questionable), it is very easy to transfer these theories back again from natural history to the history of society, and altogether too naïve to maintain that thereby these assertions have been proved as eternal natural laws of society.

Let us accept for a moment the phrase "struggle for existence," for argument's sake. The most that the animal can achieve is to *collect*; man *produces,* he prepares the means of life, in the widest sense of the words, which without him nature would not have produced. This makes impossible any unqualified transference of the laws of life in animal societies to human society. Production soon brings it about that the so-called struggle for existence no longer turns on pure means of existence, but on means of enjoyment and development. Here—where the means of development are socially produced—the categories taken from the animal kingdom are already totally inapplicable. Finally, under the capitalist mode of production, production reaches such a high level that society can no longer consume the means of life, enjoyment and development that have been produced, because for the great mass of producers access to these means is artificially and forcibly barred; and therefore every ten years a crisis restores the equilibrium by destroying not only the means of life, enjoyment and development that have been produced, but also a great part of the productive forces themselves. Hence the so-called struggle for existence assumes the form: to *protect* the products and productive forces produced by bourgeois capitalist society against the destructive, ravaging effect of this capitalist social order, by taking control of social production and distribution out of the hands of the ruling capitalist class, which has become incapable of this function, and transferring it to the producing masses—and that is the socialist revolution.

The conception of history as a series of class struggles is already much richer in content and deeper than merely reducing it to weakly distinguished phases of the struggle for existence.

Frederick Engels, *Dialectics of Nature,*
pp. 404-405.

E. The Passage from Animal Necessity to Human Freedom

The seizure of the means of production by society puts an end to commodity production, and therewith to the domination of the product over the producer. Anarchy in social production is replaced by conscious organisation on a planned basis. The struggle for individual existence comes to an end. And at this point, in a certain sense, man finally cuts himself off from the animal world, leaves the conditions of animal existence behind him and enters conditions which are really human. The conditions of existence forming man's environment, which up to now have dominated man, at this point pass under the dominion and control of man, who now for the first time becomes the real conscious master of Nature, because and in so far as he has become master of his own social organisation. The laws of his own social activity, which have hitherto confronted him as external, dominating laws of Nature, will then be applied by man with complete understanding, and hence will be dominated by man. Men's own social organisation which has hitherto stood in opposition to them as if arbitrarily decreed by Nature and history, will then become the voluntary act of men themselves. The objective, external forces which have hitherto dominated history, will then pass under the control of men themselves. It is only from this point that men, with full consciousness, will fashion their own history; it is only from this point that the social causes set in motion by men will have, predominantly and in constantly increasing measure, the effects willed by men. It is humanity's leap from the realm of necessity into the realm of freedom.

To carry through his world-emancipating act is the historical mission of the modern proletariat. And it is the task of scientific socialism, the theoretical expression of the proletarian movement, to establish the historical conditions and, with these, the nature of this act, and thus to bring to the consciousness of the now oppressed class the conditions and nature of the act which it is its destiny to accomplish.

Frederick Engels, Herr Eugen Dühring's
Revolution in Science, pp. 309-310.

F. Man's Collective Solidarity Versus Nature

Dear Mr. Lavrov,

At last, after returning from a trip to Germany, I have taken up your article[1] which I just read with great interest. The following are my observations on it, couched in German, which will permit me to be more concise.

---

1. In this letter Engels, at Lavrov's request, assesses the latter's article "Socialism and the Struggle for Existence" published in the journal *Vperyod* of September 15 (3), 1875.—Ed.

1) Of the Darwinian doctrine I accept the *theory of evolution,* but Darwin's method of proof (struggle for life, natural selection) I consider only a first, provisional, imperfect expression of a newly discovered fact. Until Darwin's time the very people who now see everywhere only *struggle* for existence (Vogt, Büchner, Moleschott, etc.) emphasized precisely *co-operation* in organic nature, the fact that the vegetable kingdom supplies oxygen and nutriment to the animal kingdom and conversely the animal kingdom supplies plants with carbonic acid and manure, which was particularly stressed by Liebig. Both conceptions are justified within certain limits, but the one is as one-sided and narrow-minded as the other. The interaction of bodies in nature—inanimate as well as animate—includes both harmony and collision, struggle and co-operation. When therefore a self-styled natural scientist takes the liberty of reducing the whole of historical development with all its wealth and variety to the one-sided and meagre phrase "struggle for existence," a phrase which even in the sphere of nature can be accepted only *cum grano salis,* such a procedure really contains its own condemnation.

2) Of the three "convinced Darwinists" you cite, only Hellwald apparently deserves mention. Seidlitz is at best only a very minor luminary and Robert Byr a novelist one of whose novels, *Dreimal,* is at present appearing in *Über Land und Meer.* That's a fitting place for his entire rodomontade.

3) I do not deny the advantages of your method of attack, which I would like to call psychological; but I would have chosen another method. Every one of us is influenced more or less by the intellectual environment in which he mostly moves. For Russia, where you know your public better than I, and for a propaganda journal that appeals to the moral sense, your method is probably the better one. For Germany, where false sentimentality has done and still does so much damage, it would not fit; it would be misunderstood, sentimentally perverted. In our country it is hatred rather than love that is needed—at least in the immediate future—and more than anything else a shedding of the last remnants of German idealism, an establishment of the material facts in their historical rights. I should therefore attack—and perhaps will when the time comes—these bourgeois Darwinists in about the following manner:

The whole Darwinist teaching of the struggle for existence is simply a transference from society to living nature of Hobbes's doctrine of *bellum omnium contra omnes* and of the bourgeois-economic doctrine of competition together with Malthus's theory of population. When this conjurer's trick has been performed (and I question its absolute permissibility, as I have indicated in point 1, particularly as far as the Malthusian theory is concerned), the same theories are transferred back again from organic nature into history and it is now claimed that their validity as eternal laws of human society has been proved. The puerility of this procedure is so obvious that not a word need be said about it. But if I wanted to go into the matter more thoroughly I should do so by depicting them in the first place as bad *economists* and only in the second place as bad naturalists and philosophers.

4) The essential difference between human and animal society consists in the fact that animals at most *collect* while men *produce*. This sole but cardinal difference alone makes it impossible simply to transfer laws of animal societies to human societies. It makes it possible, as you properly remark, "for man to struggle not only for existence but also for pleasures and *for the increase of his pleasures,* . . . to be ready to renounce his lower pleasures for the highest pleasure." Without disputing your further conclusions from this I would, proceeding from my premises, make the following inferences: At a certain stage the production of man thus attains such a high level that not only necessaries but also luxuries, at first, true enough, only for a minority, are produced. The struggle for existence—if we permit this category for the moment to be valid—is thus transformed into a struggle for pleasures, no longer for mere means of *subsistence* but for means of *development, socially produced* means of development, and to this stage the categories derived from the animal kingdom are no longer applicable. But if, as has now happened, production in its capitalist form produces a far greater quantity of means of subsistence and development than capitalist society can consume because it keeps the great mass of real producers artificially away from these means of subsistence and development; if this society is forced by its own law of life constantly to increase this output which is already too big for it and therefore periodically, every ten years, reaches the point where it destroys not only a mass of products but even productive forces—what sense is there left in all this talk of "struggle for existence"? The struggle for existence can then consist only in this: that the producing class takes over the management of production and distribution from the class that was hitherto entrusted with it but has now become incompetent to handle it, and there you have the socialist revolution.

A propos. Even the mere contemplation of previous history as a series of class struggles suffices to make clear the utter shallowness of the con-

ception of this history as a feeble variety of the "struggle for existence."
I would therefore never do this favour to these false naturalists.

5) For the same reason I would have changed accordingly the formulation of the following proposition of yours, which is essentially quite correct: that to facilitate the struggle the idea of solidarity could finally . . .
grow to a point where it will embrace all mankind and oppose it, as a
society of brothers living in solidarity, to the rest of the world—the world
of minerals, plants, and animals.

6) On the other hand I cannot agree with you that the *bellum omnium
contra omnes* was the first phase of human development. In my opinion,
the social instinct was one of the most essential levers of the evolution of
man from the ape. The first men must have lived in bands and as far as we
can peer into the past we find that this was the case.

Frederick Engels to P. L. Lavrov, November 12-17, 1875, Karl Marx
and Frederick Engels, *Selected Correspondence,* pp. 366-369.

G. Man's Brain and Thoughts Are the Products of Nature and Are in Correspondence with It

But if the further question is raised: what then are thought and consciousness, and whence they come, it becomes apparent that they are
products of the human brain and that man himself is a product of Nature,
which has been developed in and along with its environment: whence it is
self-evident that the products of the human brain, being in the last analysis
also products of Nature, do not contradict the rest of Nature but are in
correspondence with it.

Frederick Engels, *Herr Eugen Dühring's
Revolution in Science,* pp. 42-43.

H. Freedom as Insight into and Control over Natural Necessity

Hegel was the first to state correctly the relation between freedom and
necessity. To him, freedom is the appreciation of necessity. "Necessity
is *blind* only *in so far as it is not understood.*" Freedom does not consist
in the dream of independence of natural laws, but in the knowledge of
these laws, and in the possibility this gives of systematically making them
work towards definite ends. This holds good in relation both to the laws
of external nature and to those which govern the bodily and mental exis-

tence of men themselves—two classes of laws which we can separate from
each other at most only in thought but not in reality. Freedom of the
will therefore means nothing but the capacity to make decisions with real
knowledge of the subject. Therefore the *freer* a man's judgment is in relation to a definite question, with so much the greater *necessity* is the
content of this judgement determined; while the uncertainty, founded
on ignorance, which seems to make an arbitrary choice among many different and conflicting possible decisions, shows by this precisely that it
is not free, that it is controlled by the very object it should itself control.
Freedom therefore consists in the control over ourselves and over external
nature which is founded on knowledge of natural necessity; it is therefore
necessarily a product of historical development. The first men who separated themselves from the animal kingdom were in all essentials as unfree
as the animals themselves, but each step forward in civilisation was a step
towards freedom. On the threshold of human history stands the discovery
that mechanical motion can be transformed into heat: the production of
fire by friction; at the close of the development so far gone through stands
the discovery that heat can be transformed into mechanical motion: the
steam engine. And, in spite of the gigantic and liberating revolution in
the social world which the steam engine is carrying through—and which
is not yet half completed—it is beyond question that the generation of
fire by friction was of even greater effectiveness for the liberation of mankind. For the generation of fire by friction gave man for the first time
control over one of the forces of Nature, and thereby separated him for
ever from the animal kingdom. The steam engine will never bring about
such a mighty leap forward in human development, however important
it may seem in our eyes as representing all those powerful productive
forces dependent on it—forces which alone make possible a state of society
in which there are no longer class distinctions or anxiety over the means
of subsistence for the individual, and in which for the first time there can
be talk of real human freedom and of an existence in harmony with the
established laws of Nature. But how young the whole of human history
still is, and how ridiculous it would be to attempt to ascribe any absolute
validity to our present views, is evident from the simple fact that all past
history can be characterised as the history of the epoch from the practical
discovery of the transformation of mechanical motion into heat up to
that of the transformation of heat into mechanical motion.

Frederick Engels, *Herr Eugen Dühring's
Revolution in Science*, pp. 125-126.

V. Man's Application of Technology to Nature

*We now know that chimpanzees make and use simple tools, and that
the hominid Australopithecus, whose remains were found by the Leakeys
in Olduvai Gorge in Tanganyika in 1959, made and used tools some
1,750,000 years ago. Recent information places the tool-making of this
hominid at 2.5 million years ago. The extensive use of tools upon nature
sets the human species apart, and, of course, the most primitive stone
tools of apemen are the basis for all human technology. For Marx and
Engels, technology is therefore the pivot for understanding and guiding
man's relation to nature.*   —Ed.

A. The Technology of Nature and of Man

Darwin has interested us in the history of Nature's Technology, *i.e.,*
in the formation of the organs of plants and animals, which organs serve
as instruments of production for sustaining life. Does not the history of
the productive organs of man, of organs that are the material basis of all
social organisation, deserve equal attention? And would not such a history
be easier to compile, since, as Vico says, human history differs from natural
history in this, that we have made the former, but not the latter? Technology discloses man's mode of dealing with Nature, the process of production by which he sustains his life, and thereby also lays bare the mode of
formation of his social relations, and of the mental conceptions that flow
from them.

Karl Marx, *Capital*, vol. 1, p. 372 n. 3.

B. The Motive Power of Machines Is Self-Generated or Natural

All fully developed machinery consists of three essentially different
parts, the motor mechanism, the transmitting mechanism, and finally
the tool or working machine. The motor mechanism is that which puts
the whole in motion. It either generates its own motive power, like the
steam-engine, the caloric engine, the electromagnetic machine, &c., or
it receives its impulse from some already existing natural force, like the
water-wheel from a head of water, the wind-mill from wind, &c. The
transmitting mechanism, composed of fly-wheels, shafting, toothed wheels,

pullies, straps, ropes, bands, pinions, and gearing of the most varied kinds, regulates the motion, changes its form where necessary, as for instance, from linear to circular, and divides and distributes it among the working machines. These two first parts of the whole mechanism are there, solely for putting the working machines in motion, by means of which motion the subject of labour is seized upon and modified as desired. The tool or working machine is that part of the machinery with which the industrial revolution of the 18th century started. And to this day it constantly serves as such a starting-point, whenever a handicraft, or a manufacture, is turned into an industry carried on by machinery.

Karl Marx, *Capital,* vol. 1, p. 373.

C. *The Productive Power of Man's Machinery Works Gratuitously, Like the Forces of Nature*

After making allowance, both in the case of the machine and of the tool, for their average daily cost, that is for the value they transmit to the product by their average daily wear and tear, and for their consumption of auxiliary substances, such as oil, coal, and so on, they each do their work gratuitously, just like the forces furnished by Nature without the help of man. The greater the productive power of the machinery compared with that of the tool, the greater is the extent of its gratuitous service compared with that of the tool. In Modern Industry man succeeded for the first time in making the product of his past labour work on a large scale gratuitously, like the forces of Nature.

Karl Marx, *Capital,* vol. 1, p. 388.

VI. The Mutual Transformation of Man and Nature Through Labor

Marx and Engels examined in detail the transforming effects of labor upon both man and nature. As man is nourished by the materials of nature, so through the necessity of labor he shapes nature to his purposes. The earth is man's "original larder" and his "original tool house." In the process of labor, man uses as instruments of production natural things, such as soil and water, the products of his own labor, and his own labor-

*power. Through labor in and with nature, individual persons communally
satisfy their needs. It is on this basis of natural history that Marx and
Engels showed that nature and history cannot be separated in the way
that the idealists of their day and our own day have tried to do. "The
celebrated 'unity of man with nature' has always existed in industry."
Accordingly, the evolution of man's relation to nature can be traced
from the primitive "umbilical" one to a "reasonable" one through various
stages. "Historical materialism" is the logical outcome of such an analysis.*

—Ed.

A. Labor-power Is Human Energy Nourished by Matter

What Lucretius says is self-evident: "nil posse creari de nihilo," out of
nothing, nothing can be created. Creation of value is transformation of
labour-power into labour. Labour-power itself is energy transferred to a
human organism by means of nourishing matter.

Karl Marx, *Capital,* vol. 1, p. 215 n. 1.

B. Labor Combined with Natural Material Produces Use-Value

But coats and linen, like every other element of material wealth that
is not the spontaneous produce of Nature, must invariably owe their
existence to a special productive activity, exercised with a definite aim,
an activity that appropriates particular nature-given materials to particular
human wants. So far therefore as labour is a creator of use-value, is useful
labour, it is a necessary condition, independent of all forms of society, for
the existence of the human race; it is an eternal nature-imposed necessity,
without which there can be no material exchanges between man and
Nature, and therefore no life.

The use-values, coat, linen, &c., *i.e.,* the bodies of commodities, are
combinations of two elements—matter and labour. If we take away the
useful labour expended upon them, a material substratum is always left,
which is furnished by Nature without the help of man. The latter can
work only as Nature does, that is by changing the form he is constantly
helped by natural forces. We see, then, that labour is not the only source
of material wealth, of use-values produced by labour. As William Petty
puts it, labour is its father and the earth its mother.

Karl Marx, *Capital,* vol. 1, pp. 42-43.

C. The Labor-Process as the Production of Use-Values Through the Application of Man's Instruments to Nature

The capitalist buys labour-power in order to use it; and labour-power in use is labour itself. The purchaser of the labour-power consumes it by setting the seller of it to work. By working, the latter becomes actually, what before he only was potentially, labour-power in action, a labourer. In order that his labour may re-appear in a commodity, he must, before all things, expend it on something useful, on something capable of satisfying a want of some sort. Hence, what the capitalist sets the labourer to produce is a particular use-value, a specified article. The fact that the production of use-values, or goods, is carried on under the control of a capitalist and on his behalf, does not alter the general character of that production. We shall, therefore, in the first place, have to consider the labour-process independently of the particular form it assumes under given social conditions.

Labour is, in the first place, a process in which both man and Nature participate, and in which man of his own accord starts, regulates, and controls the material re-actions between himself and Nature. He opposes himself to Nature as one of her own forces, setting in motion arms and legs, head and hands, the natural forces of his body, in order to appropriate Nature's productions in a form adapted to his own wants. By thus acting on the external world and changing it, he at the same time changes his own nature. He develops his slumbering powers and compels them to act in obedience to his sway. We are not now dealing with those primitive instinctive forms of labour that remind us of the mere animal. An immeasurable interval of time separates the state of things in which a man brings his labour-power to market for sale as a commodity, from that state in which human labour was still in its first instinctive stage. We pre-suppose labour in a form that stamps it as exclusively human. A spider conducts operations that resemble those of a weaver, and a bee puts to shame many an architect in the construction of her cells. But what distinguishes the worst architect from the best of bees is this, that the architect raises his structure in imagination before he erects it in reality. At the end of every labour-process, we get a result that already existed in the imagination of the labourer at its commencement. He not only effects a change of form in the material on which he works, but he also realises a purpose of his own that gives the law to his modus operandi, and to which he must subordinate his will. And this subordination is no mere momentary act. Besides the exertion of the bodily organs, the process demands that, during the whole operation, the workman's will be steadily in consonance with his purpose. This means close attention. The less he is attracted by the nature of the work, and the mode in which it is carried on, and the less, therefore, he enjoys it as something which gives play to his bodily and mental powers, the more close his attention is forced to be.

The elementary factors of the labour-process are 1, the personal activity of man, *i.e.*, work itself, 2, the subject of that work, and 3, its instruments.

The soil (and this, economically speaking, includes water) in the virgin state in which it supplies[1] man with necessaries or the means of subsistence ready to hand, exists independently of him, and is the universal subject of human labour. All those things which labour merely separates from immediate connexion with their environment, are subjects of labour spontaneously provided by Nature. Such are fish which we catch and take from their element, water, timber which we fell in the virgin forest, and ores which we extract from their veins. If, on the other hand, the subject of labour has, so to say, been filtered through previous labour, we call it raw material; such is ore already extracted and ready for washing. All raw material is the subject of labour, but not every subject of labour is raw material; it can only become so, after it has undergone some alteration by means of labour

An instrument of labour is a thing, or a complex of things, which the labourer interposes between himself and the subject of his labour, and which serves as the conductor of his activity. He makes use of the mechanical, physical, and chemical properties of some substances in order to make other substances subservient to his aims.[2] Leaving out of consideration such ready-made means of subsistence as fruits, in gathering which a man's own limbs serve as the instruments of his labour, the first thing of which the labourer possesses himself is not the subject of labour but its

---

1. "The earth's spontaneous productions being in small quantity, and quite independent of man, appear, as it were, to be furnished by Nature, in the same way as a small sum is given to a young man, in order to put him in a way of industry, and of making his fortune." (James Steuart: "Principles of Polit. Econ." edit. Dublin, 1770. v. I, p. 116.)

2. "Reason is just as cunning as she is powerful. Her cunning consists principally in her mediating activity, which, by causing objects to act and react on each other in accordance with their own nature, in this way, without any direct interference in the process, carries out reason's intentions." (Hegel: "Enzyklopädie, Erster Theil, Die Logik," Berlin, 1840, p. 382.)

instrument. Thus Nature becomes one of the organs of his activity, one that he annexes to his own bodily organs, adding stature to himself in spite of the Bible. As the earth is his original larder, so too it is his original tool house. It supplies him, for instance, with stones for throwing, grinding, pressing, cutting, &c. The earth itself is an instrument of labour, but when used as such in agriculture implies a whole series of other instruments and a comparatively high development of labour.[3] No sooner does labour undergo the least development, than it requires specially prepared instruments. Thus in the oldest caves we find stone implements and weapons. In the earliest period of human history domesticated animals, *i.e.,* animals which have been bred for the purpose, and have undergone modifications by means of labour, play the chief part as instruments of labour along with specially prepared stones, wood, bones, and shells.[4] The use and fabrication of instruments of labour, although existing in the germ among certain species of animals, is specifically characteristic of the human labour-process, and Franklin therefore defines man as a tool-making animal. Relics of bygone instruments of labour possess the same importance for the investigation of extinct economic forms of society, as do fossil bones for the determination of extinct species of animals. It is not the articles made, but how they are made, and by what instruments, that enables us to distinguish different economic epochs.[5] Instruments of labour not only supply a standard of the degree of development to which human labour has attained, but they are also indicators of the social conditions under which that labour is carried on. Among the

---

3. In his otherwise miserable work ("Theorie de l'Econ. Polit." Paris, 1815), Ganilh enumerates in a striking manner in opposition to the "Physiocrats" the long series of previous processes necessary before agriculture properly so called can commence.

4. Turgot in his "Réflexions sur la Formation et la Distribution des Richesses" (1766) brings well into prominence the importance of domesticated animals to early civilisation.

5. The least important commodities of all for the technological comparison of different epochs of production are articles of luxury, in the strict meaning of the term. However little our written histories up to this time notice the development of material production, which is the basis of all social life, and therefore of all real history, yet prehistoric times have been classified in accordance with the results, not of so-called historical, but of materialistic investigations. These periods have been divided to correspond with the materials from which their implements and weapons were made, viz., into the stone, the bronze, and the iron ages.

instruments of labour, those of a mechanical nature, which, taken as a whole, we may call the bone and muscles of production, offer much more decided characteristics of a given epoch of production, than those which, like pipes, tubs, baskets, jars, &c., serve only to hold the materials for labour, which latter class, we may in a general way, call the vascular system of production. The latter first begins to play an important part in the chemical industries.

In a wider sense we may include among the instruments of labour, in addition to those things that are used for directly transferring labour to its subject, and which therefore, in one way or another, serve as conductor of activity, all such objects as are necessary for carrying on the labour-process. These do not enter directly into the process, but without them it is either impossible for it to take place at all, or possible only to a partial extent. Once more we find the earth to be a universal instrument of this sort, for it furnishes a locus standi to the labourer and a field of employment for his activity. Among instruments that are the result of previous labour and also belong to this class, we find workshops, canals, roads, and so forth.

In the labour-process, therefore, man's activity, with the help of the instruments of labour, effects an alteration, designed from the commencement, in the material worked upon. The process disappears in the product, the latter is a use-value. Nature's material adapted by a change of form to the wants of man. Labour has incorporated itself with its subject: the former is materialised, the latter transformed. That which in the labourer appeared as movement, now appears in the product as a fixed quality without motion. The blacksmith forges and the product is a forging.

If we examine the whole process from the point of view of its result, the product, it is plain that both the instruments and the subject of labour are means of production,[6] and that the labour itself is productive labour.[7]

Though a use-value, in the form of a product, issues from the labour-process, yet other use-values, products of previous labour, enter into it as means of production. The same use-value is both the product of a previous

6. It appears paradoxical to assert, that uncaught fish, for instance, are a means of production in the fishing industry. But hitherto no one has discovered the art of catching fish in waters that contain none.

7. This method of determining, from the standpoint of the labour-process alone, what is productive labour, is by no means directly applicable to the case of the capitalist process of production.

process, and a means of production in a later process. Products are therefore not only results, but also essential conditions of labour.

With the exception of the extractive industries, in which the material for labour is provided immediately by Nature, such as mining, hunting, fishing, and agriculture (so far as the latter is confined to breaking up virgin soil), all branches of industry manipulate raw material, objects already filtered through labour, already products of labour. Such is seed in agriculture. Animals and plants, which we are accustomed to consider as products of Nature, are in their present form, not only products of, say last year's labour, but the result of a gradual transformation, continued through many generations, under man's superintendence, and by means of his labour. But in the great majority of cases, instruments of labour show even to the most superficial observer, traces of the labour of past ages.

Raw material may either form the principal substance of a product, or it may enter into its formation only as an accessory. An accessory may be consumed by the instruments of labour, as coal under a boiler, oil by a wheel, hay by draft-horses, or it may be mixed with the raw material in order to produce some modification thereof, as chlorine into unbleached linen, coal with iron, dye-stuff with wool, or again, it may help to carry on the work itself, as in the case of the materials used for heating and lighting workshops. The distinction between principal substance and accessory vanishes in the true chemical industries, because there none of the raw material re-appears, in its original composition, in the substance of the product.[8]

Every object possesses various properties, and is thus capable of being applied to different uses. One and the same product may therefore serve as raw material in very different processes. Corn, for example, is a raw material for millers, starch-manufacturers, distillers, and cattle-breeders. It also enters as raw material into its own production in the shape of seed; coal, too, is at the same time the product of, and a means of production in, coal-mining.

Again, a particular product may be used in one and the same process, both as an instrument of labour and as raw material. Take, for instance, the fattening of cattle, where the animal is the raw material, and at the same time an instrument for the production of manure.

---

8. Storch calls true raw materials "matières," and accessory material "matériaux." Cherbuliez describes accessories as "matières instrumentales."

A product, though ready for immediate consumption, may yet serve as raw material for a further product, as grapes when they become the raw material for wine. On the other hand, labour may give us its product in such a form, that we can use it only as raw material, as is the case with cotton, thread, and yarn. Such a raw material, though itself a product, may have to go through a whole series of different processes: in each of these in turn, it serves, with constantly varying form, as raw material, until the last process of the series leaves it a perfect product, ready for individual consumption, or for use as an instrument of labour.

Hence we see, that whether a use-value is to be regarded as raw material as instrument of labour, or as product, this is determined entirely by its function in the labour-process, by the position it there occupies: as this varies, so does its character.

Whenever therefore a product enters as a means of production into a new labour-process, it thereby loses its character of product, and becomes a mere factor in the process. A spinner treats spindles only as implements for spinning, and flax only as the material that he spins. Of course it is impossible to spin without material and spindles; and therefore the existen‹ of these things as products, at the commencement of the spinning operation, must be presumed: but in the process itself, the fact that they are products of previous labour, is a matter of utter indifference; just as in the digestive process, it is of no importance whatever, that bread is the produce of the previous labour of the farmer, the miller, and the baker. On the contrary, it is generally by their imperfections as products, that the means of production in any process assert themselves in their character of products. A blunt knife or weak thread forcibly remind us of Mr. A., the cutler, or Mr. B., the spinner. In the finished product the labour by means of which it has acquired its useful qualities is not palpable, has apparently vanished.

A machine which does not serve the purposes of labour, is useless. In addition, it falls a prey to the destructive influence of natural forces. Iron rusts and wood rots. Yarn with which we neither weave nor knit, is cotton wasted. Living labour must seize upon these things and rouse them from their death-sleep, change them from mere possible use-values into real and effective ones. Bathed in the fire of labour, appropriated as part and parcel of labour's organism, and, as it were, made alive for the performance of their functions in the process, they are in truth consumed, but consume‹

with a purpose, as elementary constituents of new use-values, of new products, ever ready as means of subsistence for individual consumption, or as means of production for some new labour-process.

If then, on the one hand, finished products are not only results, but also necessary conditions, of the labour-process, on the other hand, their assumption into that process, their contact with living labour, is the sole means by which they can be made to retain their character of use-values, and be utilised.

Labour uses up its material factors, its subject and its instruments, consumes them, and is therefore a process of consumption. Such productive consumption is distinguished from individual consumption by this, that the latter uses up products, as means of subsistence for the living individual; the former, as means whereby alone, labour, the labour-power of the living individual, is enabled to act. The product, therefore, of individual consumption, is the consumer himself, the result of productive consumption, is a product distinct from the consumer.

In so far then, as its instruments and subjects are themselves products, labour consumes products in order to create products, or in other words, consumes one set of products by turning them into means of production for another set. But, just as in the beginning, the only participators in the labour-process were man and the earth, which latter exists independently of man, so even now we still employ in the process many means of production, provided directly by Nature, that do not represent any combination of natural substances with human labour.

The labour-process, resolved as above into its simple elementary factors, is human action with a view to the production of use-values, appropriation of natural substances to human requirements; it is the necessary condition for effecting exchange of matter between man and Nature; it is the everlasting Nature-imposed condition of human existence, and therefore is independent of every social phase of that existence, or rather, is common to every such phase. It was, therefore, not necessary to represent our labourer in connexion with other labourers; man and his labour on one side, Nature and its materials on the other, sufficed. As the taste of the porridge does not tell you who grew the oats, no more does this simple process tell you of itself what are the social conditions under which it is taking place, whether under the slave-owner's brutal lash, or the anxious eye of the capitalist, whether Cincinnatus carries it on in tilling his modest farm or a savage in killing wild animals with stones.[9]

Let us now return to our would-be capitalist. We left him just after he had purchased, in the open market, all the necessary factors of the labour-process; its objective factors, the means of production, as well as its sub-jective factor, labour-power. With the keen eye of an expert, he has selecte the means of production and the kind of labour-power best adapted to his particular trade, be it spinning, bootmaking, or any other kind. He then proceeds to consume the commodity, the labour-power that he has just bought, by causing the labourer, the impersonation of that labour-power, to consume the means of production by his labour. The general character of the labour-process is evidently not changed by the fact, that the labourer works for the capitalist instead of for himself; moreover, the particular methods and operations employed in bootmaking or spinning are not immediately changed by the intervention of the capitalist. He must begin by taking the labour-power as he finds it in the market, and consequently be satisfied with labour of such a kind as would be found in the period immediately preceding the rise of capitalists. Changes in the methods of production by the subordination of labour to capital, can take place only at a later period, and therefore will have to be treated of in a later chapter.

The labour-process, turned into the process by which the capitalist consumes labour-power, exhibits two characteristic phenomena. First, the labourer works under the control of the capitalist to whom his labour belongs; the capitalist taking good care that the work is done in a proper manner, and that the means of production are used with intelligence, so that there is no unnecessary waste of raw material, and no wear and tear of the implements beyond what is necessarily caused by the work.

Secondly, the product is the property of the capitalist and not that of the labourer, its immediate producer. Suppose that a capitalist pays for a day's labour-power at its value; then the right to use that power for a day belongs to him, just as much as the right to use any other commodity, such as a horse that he has hired for the day. To the purchaser of a com-

9. By a wonderful feat of logical acumen, Colonel Torrens has discovered, in this stone of the savage the origin of capital. "In the first stone which he [the savage] flings at the wild animal he pursues, in the first stick that he seizes to strike down the fruit which hangs above his reach, we see the appropriation of one article for the purpose of aiding in the acquisition of another, and thus discover the origin of capital." (R. Torrens: "An Essay on the Production of Wealth," &c., pp. 70-71.)

modity belongs its use, and the seller of labour-power, by giving his labour, does no more, in reality, than part with the use-value that he has sold. From the instant he steps into the workshop, the use-value of his labour-power, and therefore also its use, which is labour, belongs to the capitalist. By the purchase of labour-power, the capitalist incorporates labour, as a living ferment, with the lifeless constituents of the product. From his point of view, the labour-process is nothing more than the consumption of the commodity purchased, *i.e.*, of labour-power; but this consumption cannot be effected except by supplying the labour-power with the means of production. The labour-process is a process between things that the capitalist has purchased, things that have become his property. The product of this process belongs, therefore, to him, just as much as does the wine which is the product of a process of fermentation completed in his cellar.[10]

Karl Marx, *Capital*, vol. 1, pp. 177-185.

D. Natural and Human Instruments of Production

From the first, there follows the premise of a highly developed division of labour and an extensive commerce; from the second, the locality. In the first case the individuals must be brought together, in the second they find themselves alongside the given instrument of production as instruments of production themselves. Here, therefore, arises the difference between natural instruments of production and those created by civilization. The field (water, etc.) can be regarded as a natural instrument

---

10. "Products are appropriated before they are converted into capital; this conversion does not secure them from such appropriation." (Cherbuliez: "Richesse ou Pauvreté," edit. Paris, 1841, p. 54.) "The Proletarian, by selling his labour for a definite quantity of the necessaries of life, renounces all claim to a share in the product. The mode of appropriation of the products remains the same as before; it is in no way altered by the bargain we have mentioned. The product belongs exclusively to the capitalist, who supplied the raw material and the necessaries of life; and this is a rigorous consequence of the law of appropriation, a law whose fundamental principle was the very opposite, namely, that every labourer has an exclusive right to the ownership of what he produces." (l. c., p. 58.) "When the labourers receive wages for their labour . . . the capitalist is then the owner not of the capital only" (he means the means of production) "but of the labour also. If what is paid as wages is included, as it commonly is, in the term capital, it is absurd to talk of labour separately from capital. The word capital as thus employed includes labour and capital both." (James Mill: "Elements of Pol. Econ.," &c., Ed. 1821, pp. 70, 71.)

of production. In the first case, that of the natural instrument of productic
individuals are subservient to nature; in the second, to a product of labour.
In the first case, therefore, property (landed property) appears as direct
natural domination, in the second as domination of labor, particularly of
accumulated labour, capital. The first case presupposes that the individuals
are united by some bond, family, tribe, the land itself, etc.; the second that
they are independent of one another and are only held together by exchange. In the first case, what is involved is chiefly an exchange between
men and nature, in which the labour of the former is exchanged for the
products of the latter; in the second, it is predominantly an exchange of
men among themselves. In the first case, average human common-sense is
adequate—physical and mental activity are as yet not separated at all; in
the second, the division between physical and mental labour must already
be practically completed. In the first case, the domination of the proprieto
over the propertyless may be based on a personal relationship, on a kind
of community; in the second, it must have taken on a material shape in
a third party—money. In the first case, small industry exists, but determined by the utilization of the natural instrument of production and
therefore without the distribution of labour among various individuals;
in the second, industry exists only in and through the division of labour.

Karl Marx and Friedrich Engels,
The German Ideology, pp. 63-64.

E. Communism Unites Men and Nature

Communism differs from all previous movements in that it overturns
the basis of all earlier relations of production and intercourse, and for
the first time consciously treats all natural premises as the creatures of
men, strips them of their natural character and subjugates them to the
power of individuals united. Its organization is, therefore, essentially
economic, the material production of the conditions of this unity; it
turns existing conditions into conditions of unity. The reality, which
communism is creating, is precisely the real basis for rendering it impossible that anything should exist independently of individuals, in so far as
things are only a product of the preceding intercourse of individuals themselves. Thus the communists in practice treat the conditions created by
production and intercourse as inorganic conditions, without, however,
imagining that it was the plan or the destiny of previous generations to give

them material, and without believing that these conditions were inorganic for the individuals creating them.

Karl Marx and Friedrich Engels,
The German Ideology, p. 70.

F. History and Nature Are Not Separable

In the whole conception of history up to the present this real basis of history has either been totally neglected or else considered as a minor matter quite irrelevant to the course of history. History must therefore always be written according to an extraneous standard; the real production of life seems to be beyond history, while the truly historical appears to be separated from ordinary life, something extra-superterrestrial. With this the relation of man to nature is excluded from history and hence the antithesis of nature and history is created. The exponents of this conception of history have consequently only been able to see in history the political actions of princes and States, religious and all sorts of theoretical struggles, and in particular in each historical epoch have had to share the *illusion of that epoch.* For instance, if an epoch imagines itself to be actuated by purely "political" or "religious" motives, although "religion" and "politics" are only forms of its true motives, the historian accepts this opinion. The "idea," the "conception" of these conditioned men about their real practice, is transformed into the sole determining, active force, which controls and determines their practice. When the crude form in which the division of labour appears with the Indians and Egyptians calls forth the caste-system in their State and religion, the historian believes that the caste-system is the power which has produced this crude social form. While the French and the English at least hold by the political illusion, which is moderately close to reality, the Germans move in the realm of the "pure spirit," and make religious illusion the driving force of history.

Karl Marx and Friedrich Engels,
The German Ideology, p. 30.

G. The Unity and Struggle of Man with Nature Through Industry

Feuerbach's "interpretation" of the sensuous world is confined on the one hand to mere contemplation of it, and on the other to mere feeling;

he says "man" instead of "real, historical men." "Man" is really "the German." In the first case, the contemplation of the sensuous world, he necessarily lights on things which contradict his consciousness and feeling, which upset the harmony of all parts of the sensuous world and especially of man and nature, a harmony he presupposes.[1] To push these on one side, he must take refuge in a double perception, a profane one which only perceives the "flatly obvious" and a higher more philosophical one which perceives the "true essence" of things. He does not see how the sensuous world around him is, not a thing given direct from all eternity, ever the same, but the product of industry and of the state of society; and, indeed, in the sense that it is an historical product, the result of the activity of a whole succession of generations, each standing on the shoulders of the preceding one, developing its industry and its intercourse, modifying its social organization according to the changed needs. Even the objects of the simplest "sensuous certainty" are only given him through social development, industry and commercial intercourse. The cherry-tree, like almost all fruit-trees, was, as is well known, only a few centuries ago transplanted by commerce into our zone, and therefore only by this action of a definite society in a definite age provided for the evidence of Feuerbach's "senses." Actually, when we conceive things thus, as they really are and happened, every profound philosophical problem is resolved, as will be seen even more clearly later, quite simply into an empirical fact.

For instance, the important question of the relation of man to nature (Bruno goes so far as to speak of "the antitheses in nature and history," as though these were two separate "things" and man did not always have before him an historical nature and a natural history) out of which all the "unfathomably lofty works" on "substance" and "self-consciousness" were born, crumbles of itself when we understand that the celebrated "unity of man with nature" has always existed in industry and has existed in varying forms in every epoch according to the lesser or greater development of industry, just like the "struggle" of man with nature, right up to the development of his productive powers on a corresponding

---

1. Feuerbach's failing is not that he subordinates the flatly obvious, the sensuous appearance, to the sensuous reality established by more accurate investigation of the sensuous facts, but that he cannot in the last resort cope with the sensuous world except by looking at it with the "eyes" i.e. through the "spectacles" of the *philosopher*.

basis. Industry and commerce, production and the exchange of the necessities of life, themselves determine distribution, the structure of the different social classes and are, in turn, determined by these as to the mode
in which they are carried on; and so it happens that in Manchester, for
instance, Feuerbach sees only factories and machines where a hundred
years ago only spinning-wheels and weaving-looms were to be seen, or in
the Campagna of Rome he finds only pasture lands and swamps, where in
the time of Augustus he would have found nothing but the vineyards and
villas of Roman capitalists. Feuerbach speaks in particular of the perception of natural science; he mentions secrets which are disclosed only to
the eye of the physicist and chemist: but where would natural science be
without industry and commerce? Even this "pure" natural science is provided with an aim, as with its material, only through trade and industry,
through the sensuous activity of men. So much is this activity, this un-
ceasing sensuous labour and creation, this production, the basis of the
whole sensuous world as it now exists, that, were it interrupted only for
a year, Feuerbach would not only find an enormous change in the natural
world, but would very soon find that the whole world of men and his own
perceptive faculty, nay his own existence, were missing.

Of course, in all this the priority of external nature remains unassailed,
and all this has no application to the original men produced by "generatio
aequivoca" (spontaneous generation); but this differentiation has meaning only in so far as man is considered to be distinct from nature. For that
matter, nature, the nature that preceded human history, is not by any
means the nature in which Feuerbach lives, nor the nature which to-day
no longer exists anywhere (except perhaps on a few Australian coral-islands
of recent origin) and which, therefore, does not exist for Feuerbach. . . .

Certainly Feuerbach has a great advantage over the "pure" materialists
in that he realizes how man too is an "object of the senses." But apart
from the fact that he only conceives him as a "sensuous object," not as
"sensuous activity," because he still remains in the realm of theory and
conceives of men not in their given social connection, not under their
existing conditions of life, which have made them what they are, he never
arrives at the really existing active men, but stops at the abstraction "man,"
and gets no further than recognizing "the true, individual, corporeal man"
emotionally, i.e. he knows no other "human relationships" "of man to
man" than love and friendship, and even then idealized. He gives no
criticism of the present conditions of life. Thus he never manages to

conceive the sensuous world as the total living sensuous activity of the individuals composing it; and therefore when, for example, he sees instead of healthy men a crowd of scrofulous, over-worked and consumptive starvelings, he is compelled to take refuge in the "higher perception" and in the ideal "compensation in the species," and thus to relapse into idealism at the very point where the communist materialist sees the necessity, and at the same time the condition, of a transformation both of industry and of the social structure.

As far as Feuerbach is a materialist he does not deal with history, and as far as he considers history he is not a materialist. With him materialism and history diverge completely, a fact which explains itself from what has been said.

Karl Marx and Friedrich Engels,
The German Ideology, pp. 34-38.

H. The Evolution of Man's Relation to Nature from Worship to Reason

The religious world is but the reflex of the real world. And for a society based upon the production of commodities, in which the producers in general enter into social relations with one another by treating their products as commodities and values, whereby they reduce their individual private labour to the standard of homogeneous human labour—for such a society, Christianity with its *cultus* of abstract man, more especially in its bourgeois developments, Protestantism, Deism, &c., is the most fitting form of religion. In the ancient Asiatic and other ancient modes of production, we find that the conversion of products into commodities, and therefore the conversion of men into producers of commodities, holds a subordinate place, which, however, increases in importance as the primitive communities approach nearer and nearer to their dissolution. Trading nations, properly so called, exist in the ancient world only in its interstices, like the gods of Epicurus in the Intermundia, or like Jews in the pores of Polish society. Those ancient social organisms of production are, as compared with bourgeois society, extremely simple and transparent. But they are founded either on the immature development of man individually, who has not yet severed the umbilical cord that unites him with his fellowmen in a primitive tribal community, or upon direct relations of subjection. They can arise and exist only when the development of the productive power of labour has not risen beyond a low stage, and when, therefore, the social relations within the sphere of material life, between man and man, and between man and Nature, are correspondingly narrow. This narrowness is reflected in the ancient worship of Nature, and in the other elements of the popular religions. The religious reflex of the real world can, in any case, only then finally vanish, when the practical relations of everyday life offer to man none but perfectly intelligible and reasonable relations with regard to his fellowmen and to Nature.

Karl Marx, *Capital*, vol. 1, p. 79.

I. Stages in Man's Development in Relation to Nature

Since we are dealing with the Germans, who do not postulate anything, we must begin by stating the first premise of all human existence, and therefore of all history, the premise namely that men must be in a position to live in order to be able to "make history." But life involves before everything else eating and drinking, a habitation, clothing and many other things. The first historical act is thus the production of the means to satisfy these needs, the production of material life itself. And indeed this is an historical act, a fundamental condition of all history, which to-day, as thousands of years ago, must daily and hourly be fulfilled merely in order to sustain human life. Even when the sensuous world is reduced to a minimum, to a stick as with Saint Bruno, it presupposes the action of producing the stick. The first necessity therefore in any theory of history is to observe this fundamental fact in all its significance and all its implications and to accord it its due importance. This, as is notorious, the Germans have never done, and they have never therefore had an earthly basis for history and consequently never a historian. The French and the English, even if they have conceived the relation of this fact with so-called history only in an extremely one-sided fashion, particularly as long as they remained in the toils of political ideology, have nevertheless made the first attempts to give the writing of history a materialistic basis by being the first to write histories of civil society, of commerce and industry.

The second fundamental point is that as soon as a need is satisfied, (which implies the action of satisfying, and the acquisition of an instrument), new needs are made; and this production of new needs is the first historical act. Here we recognize immediately the spiritual ancestry of the great historical wisdom of the Germans who, when they run out of positive material and when they can serve up neither theological nor political nor literary rubbish, do not write history at all, but invent the "prehistoric era." They do not, however, enlighten us as to how we proceed from this nonsensical "prehistory" to history proper; although, on the other hand, in their historical speculation they seize upon this "prehistory" with especial eagerness because they imagine themselves safe there from interference on the part of "crude facts," and, at the same time, because there they can give full rein to their speculative impulse and set up and knock down hypotheses by the thousand.

The third circumstance which, from the very first, enters into historical development, is that men, who daily remake their own life, begin to make other men, to propagate their kind: the relation between man and wife, parents and children, the FAMILY. The family which to begin with is the only social relationship, becomes later, when increased needs create new social relations and the increased population new needs, a subordinate one (except in Germany), and must then be treated and analysed according to the existing empirical data,[1] not according to "the concept of the family," as is the custom in Germany. These three aspects of social activity are not of course to be taken as three different stages, but just, as I have said, as three aspects or, to make it clear to the Germans, three "moments," which have existed simultaneously since the dawn of history and the first men, and still assert themselves in history to-day.

The production of life, both of one's own in labour and of fresh life in

---

1. The building of houses. With savages each family has of course its own cave or hut like the separate family tent of the nomads. This separate domestic economy is made only the more necessary by the further development of private property. With the agricultural peoples a communal domestic economy is just as impossible as a communal cultivation of the soil. A great advance was the building of towns. In all previous periods, however, the abolition of individual economy, which is inseparable from the abolition of private property, was impossible for the simple reason that the material conditions governing it were not present. The setting-up of a communal domestic economy presupposes the development of machinery, of the use of natural forces and of many other productive forces—e.g. of water-supplies, of gas-lighting, steam-heating, etc., the removal of the antagonism of town and country. Without these conditions a communal economy would not in itself form a new productive force; lacking any material basis and resting on a purely theoretical foundation, it would be a mere freak and would end in nothing more than a monastic economy.— What was possible can be seen in the formation of towns and the erection of communal buildings for various definite purposes (prisons, barracks, etc.). That the abolition of individual economy is inseparable from the abolition of the family is self-evident.

procreation, now appears as a double relationship: on the one hand as a natural, on the other as a social relationship. By social we understand the co-operation of several individuals, no matter under what conditions, in what manner and to what end. It follows from this that a certain mode of production, or industrial stage, is always combined with a certain mode of co-operation, or social stage, and this mode of co-operation is itself a "productive force." Further, that the multitude of productive forces accessible to men determines the nature of society, hence that the "history of humanity" must always be studied and treated in relation to the history of industry and exchange. But it is also clear how in Germany it is impossible to write this sort of history, because the Germans lack not only the necessary power of comprehension and the material but also the "evidence of their senses," for across the Rhine you cannot have any experience of these things since history has stopped happening. Thus it is quite obvious from the start that there exists a materialistic connection of men with one another, which is determined by their needs and their mode of production, and which is as old as men themselves. This connection is ever taking on new forms, and thus presents a "history" independently of the existence of any political or religious nonsense which would hold men together on its own.

Only now, after having considered four moments, four aspects of the fundamental historical relationships, do we find that man also possesses "consciousness"; but, even so, not inherent, not "pure" consciousness. From the start the "spirit" is afflicted with the curse of being "burdened" with matter, which here makes its appearance in the form of agitated layers of air, sounds, in short of language. Language is as old as consciousness, language is practical consciousness, as it exists for other men, and for that reason is really beginning to exist for me personally as well; for language, like consciousness, only arises from the need, the necessity, of intercourse with other men. Where there exists a relationship, it exists for me: the animal has no "relations" with anything, cannot have any. For the animal, its relation to others does not exist as a relation. Consciousness is therefore from the very beginning a social product, and remains so as long as men exist at all. Consciousness is at first, of course, merely consciousness concerning the immediate sensuous environment and consciousness of the limited connection with other persons and things outside the individual who is growing self-conscious. At the same time it is consciousness of nature, which first appears to men as a completely alien, all-powerful and unassailable force, with which men's relations are purely animal and by which they are overawed like beasts; it is thus a purely animal consciousness of nature (natural religion).

We see here immediately: this natural religion or animal behaviour towards nature is determined by the form of society and *vice versa.* Here, as everywhere, the identity of nature and man appears in such a way that the restricted relation of men to nature determines their restricted relation to one another, and their restricted relation to one another determines men's restricted relation to nature, just because nature is as yet hardly modified historically; and, on the other hand, man's consciousness of the necessity of associating with the individuals around him is the beginning of the consciousness that he is living in society at all. This beginning is as animal as social life itself at this stage. It is mere herd-consciousness, and at this point man is only distinguished from sheep by the fact that with him consciousness takes the place of instinct or that his instinct is a conscious one.

Karl Marx and Friedrich Engels,
The German Ideology, pp. 16-20.

VII. Precapitalist Relations of Man to Nature

Marx's investigation of the history of man's relation to nature through communal labor led to his concentration on the modes of production and distribution in precapitalist societies, as contrasted with what occurs in capitalist societies.   —Ed.

A. The Unity of Man with Man and Man with Nature

A presupposition of wage labour, and one of the historic preconditions for capital, is free labour and the exchange of this free labour for money, in order to reproduce and to realize money, to consume the use value of labour not for individual consumption, but as use value for money. Another presupposition is the separation of free labour from the objective conditions of its realization—from the means of labour and the material for labour. Thus, above all, release of the worker from the soil as his natural workshop—hence dissolution of small, free landed property as well as of communal landownership resting on the oriental commune. In both forms, the worker relates to the objective conditions of his labour as to his property; this is the natural unity of labour with its material [*sachlich*] presuppositions. The worker thus has an objective existence independent of labour. The individual relates to himself as proprietor, as master of the conditions of his reality. He relates to the others in the same way and—depending on whether this presupposition is posited as proceeding from the community or from the individual families which constitute the commune—he relates to the others as co-proprietors, as so many incarnations of the common property, or as independent proprietors like himself, independent private proprietors—besides whom the previously all-absorbing and all-predominant communal property is itself posited as a particular *ager publicus* alongside the many private landowners.

Karl Marx, *Grundrisse,* p. 471.

B. Communal Labor Appropriates Nature as Common Property

In the first form of this landed property, an initial, naturally arisen spontaneous [*naturwüchsiges*] community appears as first presupposition. Family, and the family extended as a clan [*Stamm*] , or through intermarriage between families, or combination of clans. Since we may assume that *pastoralism,* or more generally a *migratory* form of life, was the first form of the mode of existence, not that the clan settles in a specific site, but that it grazes off what it finds—humankind is not settlement-prone by nature (except possibly in a natural environment so especially fertile that they sit like monkeys on a tree; else roaming like the animals)—then the *clan community,* the natural community, appears not as a *result* of, but as a *presupposition for the communal appropriation* (temporary) *and utilization of the land.* When they finally do settle down, the extent to which this original community is modified will depend on various external, climatic, geographic, physical etc. conditions as well as on their particular natural predisposition—their clan character. This naturally arisen clan community, or, if one will, pastoral society, is the first presupposition—the communality [*Gemeinschaftlichkeit*] of blood, language customs—for the *appropriation of the objective conditions* of their life, and of their life's reproducing and objectifying activity (activity as herdsmen, hunters, tillers, etc.). The earth is the great workshop, the arsenal which furnishes both means and material of labour, as well as the seat, the *base* of the community. They relate naively to it as the *property of the community,* of the community producing and reproducing itself in living labour. Each individual conducts himself only as a link, as a member of this community as *proprietor* or *possessor.*

Karl Marx, Grundrisse, p. 472.

C. The Town as the Center of Community

The second form—and like the first it has essential modifications brought about locally, historically etc.—product of more active, historic life, of the fates and modifications of the original clans—also assumes the *community* as its first presupposition, but not, as in the first case, as the substance of which the individuals are mere accidents, or of which they form purely natural component parts—it presupposes as base not the countryside, but the town as an already created seat (centre) of the rural population (owners of land). The cultivated field here appears as a *territorium* belonging to the town; not the village as mere accessory to the land. The earth in itself—regardless of the obstacles it may place in the way of working it, really appropriating it—offers no resistance to [attempts to] relate to it as the inorganic nature of the living individual, as his workshop, as the means and object of labour and the means of life for the subject. The difficulties which the commune encounters can arise only from other communes, which have either previously occupied the land and soil, or which disturb the commune in its own occupation. War is therefore the great comprehensive task, the great communal labour which is required either to occupy the objective conditions of being there alive, or to protect and perpetuate the occupation. Hence the commune consisting of families initially organized in a warlike way—as a system of war and army, and this is one of the conditions of its being there as proprietor.

Karl Marx, Grundrisse, p. 474.

D. Appropriation of the Earth by the Commune

The main point here is this: In all these forms—in which landed property and agriculture form the basis of the economic order, and where the economic aim is hence the production of use values, i.e. the *reproduction of the individual* within the specific relation to the commune in which he is its basis—there is to be found: (1) Appropriation not through labour, but presupposed to labour; appropriation of the natural conditions of labour, of the *earth* as the original instrument of labour as well as its workshop and repository of raw materials. The individual relates simply to the objective conditions of labour as being his; [relates] to them as the inorganic nature of his subjectivity in which the latter realizes itself; the chief objective condition of labour does not itself appear as a *product* of labour, but is already there as *nature*; on one side the living individual, on the other the earth, as the objective condition of his reproduction; (2) but this *relation* to land and soil, to the earth, as the property of the labouring individual—who thus appears from the outset not merely as labouring individual, in this abstraction, but who has an *objective mode of existence* in his ownership of the land, an existence *presupposed* to his activity, and not merely as a result of it, a presupposition of his activity just like his skin, his sense organs, which of course he also reproduces and develops etc. in the life process, but which are nevertheless presuppositions of this process of his reproduction—is instantly mediated by the naturally arisen, spontaneous, more or less historically developed and modified presence of the individual as *member of a commune*—his naturally arisen presence as member of a tribe etc. An isolated individual could no more have property in land and soil than he could speak. He could, of course, live off it as substance, as do the animals. The relation to the earth as property is always mediated through the occupation of the land and soil, peacefully or violently, by the tribe, the commune, in some more or less naturally arisen or already historically developed form.

Karl Marx, *Grundrisse,* p. 485.

VIII. Capitalist Pollution and the Ruination of Nature

The development of primitive societies into class societies and eventually into capitalist society has produced a transformation in man's relation to nature. Under capitalism this relation is determined primarily by the ruling class of capitalists. Marx and Engels undertook a thorough examination of this new relation. Like the relation that they bear to workers, the relation of capitalists to nature is marked by exploitation, pollution, and ruination. Marx and Engels observed that the capitalists appropriate the resources of the earth without any cost to themselves, that they transform the earth into an "object of huckstering," that under capitalism the original unity of man and nature is breached as well as the unity of manufacture and agriculture, that man's precipitous alteration of nature issues in un-

*foreseen and harmful consequences, and that the capitalists' wastage and
exhaustion of the soil, deforestation, disruption of nature's cycle of matter,
greedy policy toward nature, and neglect of man's welfare are ruinous to
both nature and man.*   —Ed.

A. The Productive Forces of Man, Science, and Nature Cost Capital Nothing

We saw that the productive forces resulting from co-operation and division of labour cost capital nothing. They are natural forces of social labour.
So also physical forces, like steam, water, &c., when appropriated to
productive processes, cost nothing. But just as a man requires lungs to
breathe with, so he requires something that is work of man's hand, in order
to consume physical forces productively. A water-wheel is necessary to
exploit the force of water, and a steam-engine to exploit the elasticity of
steam. Once discovered, the law of the deviation of the magnetic needle
in the field of an electric current, or the law of the magnetisation of iron,
around which an electric current circulates, cost never a penny.[1] But the
exploitation of these laws for the purposes of telegraphy, &c., necessitates
a costly and extensive apparatus. The tool, as we have seen, is not exter-
minated by the machine. From being a dwarf implement of the human
organism, it expands and multiplies into the implement of a mechanism
created by man. Capital now sets the labourer to work, not with a manual
tool, but with a machine which itself handles the tools. Although, therefore, it is clear at the first glance that, by incorporating both stupendous
physical forces, and the natural sciences, with the process of production,
Modern Industry raises the productiveness of labour to an extraordinary
degree, it is by no means equally clear, that this increased productive force
is not, on the other hand, purchased by an increased expenditure of labour.
Machinery, like every other component of constant capital, creates no
new value, but yields up its own value to the product that it serves to
beget. In so far as the machine has value, and, in consequence, parts with

---

1. Science, generally speaking, costs the capitalist nothing, a fact that by no
means hinders him from exploiting it. The science of others is as much annexed
by capital as the labour of others. Capitalistic appropriation and personal appropriation, whether of science or of material wealth, are however, totally different things. Dr. Ure himself deplores the gross ignorance of mechanical science
existing among his dear machinery-exploiting manufacturers, and Liebig can a tale
unfold about the astounding ignorance of chemistry displayed by English chemical
manufacturers.

value to the product, it forms an element in the value of that product. Instead of being cheapened, the product is made dearer in proportion to the value of the machine. And it is clear as noon-day, that machines and systems of machinery, the characteristic instruments of labour of Modern Industry, are incomparably more loaded with value than the implements used in handicrafts and manufactures.

In the first place, it must be observed that the machinery, while always entering as a whole into the labour-process, enters into the value-begetting process only by bits. It never adds more value than it loses, on an average, by wear and tear. Hence there is a great difference between the value of a machine, and the value transferred in a given time by that machine to the product. The longer the life of the machine in the labour-process, the greater is that difference. It is true, no doubt, as we have already seen, that every instrument of labour enters as a whole into the labour-process, and only piece-meal, proportionally to its average daily loss by wear and tear, into the value-begetting process. But this difference between the instrument as a whole and its daily wear and tear, is much greater in a machine than in a tool, because the machine, being made from more durable material, has a longer life; because its employment, being regulated by strictly scientific laws, allows of greater economy in the wear and tear of its parts, and in the materials it consumes; and lastly, because its field of production is incomparably larger than that of a tool.

                                   Karl Marx, *Capital,* vol. 1, pp. 386-388.

In the extractive industries, mines, &c., the raw materials form no part of the capital advanced. The subject of labour is in this case not a product of previous labour, but is furnished by Nature gratis, as in the case of metals, minerals, coal, stone, &c.

                                   Karl Marx, *Capital,* vol. 1, p. 603.

Natural elements entering as agents into production, and which cost nothing, no matter what role they play in production, do not enter as components of capital, but as a free gift of Nature to capital, that is, as a free gift of Nature's productive power to labour, which, however, appears as the productiveness of capital, as all other productivity under the capitalist mode of production.

                                   Karl Marx, *Capital,* vol. 3, p. 745.

While productive labour is changing the means of production into constituent elements of a new product, their value undergoes a metempsychosis. It deserts the consumed body, to occupy the newly created one. But this transmigration takes place, as it were, behind the back of the labourer. He is unable to add new labour, to create new value, without at the same time preserving old values, and this, because the labour he adds must be of a specific useful kind; and he cannot do work of a useful kind, without employing products as the means of production of a new product, and thereby transferring their value to the new product. The property therefore which labour-power in action, living labour, possesses of preserving value, at the same time that it adds it, is a gift of Nature which costs the labourer nothing, but which is very advantageous to the capitalist inasmuch as it preserves the existing value of his capital. So long as trade is good, the capitalist is too much absorbed in money-grubbing to take notice of this gratuitous gift of labour. A violent interruption of the labour-process by a crisis, makes him sensitively aware of it.

Karl Marx, *Capital,* vol. 1, pp. 206-207.

B. Capitalist Huckstering of People and Nature

The axioms which qualify as robbery the landowner's method of deriving an income—namely, that each has a right to the product of his labor, or that no one shall reap where he has not sown—are not something *we* have invented. The first excludes the duty of feeding children; the second deprives each generation of the right to live, since each generation starts with what it inherits from the preceding generation. These axioms are, rather, implications of private property. Each one implements its implications or one abandons private property as a premise.

Indeed, the original act of appropriation itself is justified by the assertion of the still earlier *common* property right. Thus, wherever we turn, private property leads us into contradictions.

To make earth an object of huckstering—the earth which is our one and all, the first condition of our existence—was the last step toward making oneself an object of huckstering. It was and is to this very day an immorality surpassed only by the immorality of self-alienation. And the original appropriation—the monopolization of the earth by a few, the exclusion of the rest from that which is the condition of their life—yields nothing in immorality to the subsequent huckstering of the earth.

Frederick Engels, *Outlines of a Critique of Political Economy,* in Karl Marx, *The Economic and Philosophic Manuscripts of 1844,* p. 210.

C. The Separation of Man from Nature under Capitalism

The original conditions of production (or, what is the same, the reproduction of a growing number of human beings through the natural process
between the sexes; for this reproduction although it appears as appropriation of the objects by the subjects in one respect, appears in another respect also as formation, subjugation of the objects to a subjective purpose:
their transformation into results and repositories of subjective activity)
cannot themselves originally *be products*—results of production. It is not
the *unity* of living and active humanity with the natural, inorganic conditions of their metabolic exchange with nature, and hence their appropriation of nature, which requires explanation or is the result of a historic process, but rather the *separation* between these inorganic conditions of human existence and this active existence, a separation which is completely
poisited only in the relation of wage labour and capital. In the relations of
slavery and serfdom this separation does not take place; rather, one part of
society is treated by the other as itself merely an *inorganic and natural* condition of its own reproduction. The slave stands in no relation whatsoever
to the objective conditions of his labour; rather, *labour* itself, both in the
form of the slave and in that of the serf, is classified as *an inorganic condition* of production along with other natural beings, such as cattle, as an accessory of the earth. In other words: the original conditions of production
appear as natural presuppositions, *natural conditions of the producer's
existence* just as his living body, even though he reproduces and develops
it, is originally not posited by himself, but appears as the *presupposition
of his self*; his own (bodily)being is a natural presupposition, which he has
not posited. These *natural conditions of existence*, to which he relates as
to his own inorganic body, are themselves double: (1) of a subjective and
(2) of an objective nature. He finds himself a member of a family, clan,
tribe, etc.—which then, in a historic process of intermixture and antithesis
with others, takes on a different shape; and, as such a member, he relates
to a specific nature (say, here, still earth, land, soil) as his own inorganic
being, as a condition of his production and reproduction.

Karl Marx, *Grundrisse*, pp. 489-490.

D. Capitalism Rends the Unity of Agriculture and Industry, Disturbs Man's Relation to the Soil, Wastes Workers, and Robs Laborers and Soil

In the sphere of agriculture, modern industry has a more revolutionary
effect than elsewhere, for this reason, that it annihilates the peasant, that

bulwark of the old society, and replaces him by the wage-labourer. Thus the desire for social changes, and the class antagonisms are brought to the same level in the country as in the towns. The irrational, old-fashioned methods of agriculture are replaced by scientific ones. Capitalist production completely tears asunder the old bond of union which held together agriculture and manufacture in their infancy. But at the same time it creates the material conditions for a higher synthesis in the future, viz., the union of agriculture and industry on the basis of the more perfected forms they have each acquired during their temporary separation. Capitalist production, by collecting the population in great centres, and causing an ever-increasing preponderance of town population, on the one hand concentrates the historical motive power of society; on the other hand, it disturbs the circulation of matter between man and the soil, *i.e.*, prevents the return to the soil of its elements consumed by man in the form of food and clothing; it therefore violates the conditions necessary to lasting fertility of the soil. By this action it destroys at the same time the health of the town labourer and the intellectual life of the rural labourer. But while upsetting the naturally grown conditions for the maintenance of that circulation of matter, it imperiously calls for its restoration as a system, as a regulating law of social production, and under a form appropriate to the full development of the human race. In agriculture as in manufacture, the transformation of production under the sway of capital, means, at the same time, the martyrdom of the producer; the instrument of labour becomes the means of enslaving, exploiting, and impoverishing the labourer; the social combination and organisation of labour-processes is turned into an organised mode of crushing out the workman's individual vitality, freedom, and independence. The dispersion of the rural labourers over larger areas breaks their power of resistance while concentration increases that of the town operatives. In modern agriculture, as in the urban industries, the increased productiveness and quantity of the labour set in motion are bought at the cost of laying waste and consuming by disease labour-power itself. Moreover, all progress in capitalistic agriculture is a progress in the art, not only of robbing the labourer, but of robbing the soil; all progress in increasing the fertility of the soil for a given time, is a progress towards ruining the lasting sources of that fertility. The more a country starts its development on the foundation of modern industry, like the United States, for example, the more rapid is this process of destruction.[1] Capitalist production, therefore, develops technology, and the combination together of various processes into a social whole, only by sapping the original sources of all wealth—the soil and the labourer.

Karl Marx, *Capital*, vol. 1, pp. 505-507.

E. Capitalist Failure to Utilize the Waste Products of Industry, Agriculture, and Human Consumption, and Ways of Utilizing Waste

The capitalist mode of production extends the utilisation of the excretions of production and consumption. By the former we mean the waste

---

1. See Liebig: "Die Chemie in ihrer Anwendung auf Agricultur und Physiologie," 7. Auflage, 1862, and especially the "Einleitung in die Naturgesetze des Feldbaus," in the 1st Volume. To have developed from the point of view of natural science, the negative, *i.e.,* destructive side of modern agriculture, is one of Liebig's immortal merits. His summary, too, of the history of agriculture, although not free from gross errors, contains flashes of light. It is, however, to be regretted that he ventures on such haphazard assertions as the following: "By greater pulverising and more frequent ploughing, the circulation of air in the interior of porous soil is aided, and the surface exposed to the action of the atmosphere is increased and renewed; but it is easily seen that the increased yield of the land cannot be proportional to the labour spent on that land, but increases in a much smaller proportion. This law," adds Liebig, "was first enunciated by John Stuart Mill in his 'Principles of Pol. Econ.,' Vol. 1, p. 17 as follows: 'That the produce of land increases, *caeteris paribus,* in a diminishing ratio to the increase of the labourers employed' (Mill here introduces in an erroneous form the law enunciated by Ricardo's school, for since the 'decrease of the labourers employed,' kept even pace in England with the advance of agriculture, the law discovered in, and applied to, England, could have no application to that country, at all events), 'is the universal law of agricultural industry.' This is very remarkable, since Mill was ignorant of the reason for this law." (Liebig, 1. c., Bd. I., p. 143 and Note.) Apart from Liebig's wrong interpretation of the word "labour," by which word he understands something quite different from what Political Economy does, it is, in any case, "very remarkable" that he should make Mr. John Stuart Mill the first propounder of a theory which was first published by James Anderson in A. Smith's days, and was repeated in various works down to the beginning of the 19th century; a theory which Malthus, that master in plagiarism (the whole of his population theory is a shameless plagiarism), appropriated to himself in 1815; which West developed at the same time as, and independently of, Anderson; which in the year 1817 was connected by Ricardo with the general theory of value, then made the round of the world as Ricardo's theory, and in 1820 was vulgarised by James Mill, the father of John Stuart Mill; and which, finally, was reproduced by John Stuart Mill and others, as a dogma already quite commonplace, and known to every schoolboy. It cannot be denied that John Stuart Mill owes his, at all events, "remarkable" authority almost entirely to such *quid-pro-quos.*

of industry and agriculture, and by the latter partly the excretions produced by the natural exchange of matter in the human body and partly
the form of objects that remains after their consumption. In the chemical
industry, for instance, excretions of production are such by-products as
are wasted in production on a smaller scale; iron filings accumulating in
the manufacture of machinery and returning into the production of iron
as raw material, etc. Excretions of consumption are the natural waste
matter discharged by the human body, remains of clothing in the form
of rags, etc. Excretions of consumption are of the greatest importance
for agriculture. So far as their utilisation is concerned, there is an enormous
waste of them in the capitalist economy. In London, for instance, they
find no better use for the excretion of four and half million human beings
than to contaminate the Thames with it at heavy expense.

Rising prices of raw materials naturally stimulate the utilisation of
waste products.

The general requirements for the re-employment of these excretions
are: large quantities of such waste, such as are available only in large-scale
production; improved machinery whereby materials, formerly useless in
their prevailing form, are put into a state fit for new production; scientific
progress, particularly of chemistry, which reveals the useful properties of
such waste. It is true that great savings of this sort are also observed in
small scale agriculture, as prevails in, say, Lombardy, southern China,
and Japan. But on the whole, the productivity of agriculture under this
system obtains from the prodigal use of human labour-power, which is
withheld from other spheres of production.

The so-called waste plays an important role in almost every industry.
Thus, the Factory Report for December 1863 mentions as one of the
principal reasons why the English and many of the Irish farmers do not
like to grow flax, or do so but rarely, "the great waste . . . which has taken
place at the little water scutch mills . . . the waste in cotton is comparatively small, but in flax very large. The efficiency of water steeping and of
good machine scutching will reduce this disadvantage very considerably . . . .
Flax, scutched in Ireland in a most shameful way, and a large percentage
actually lost by it, equal to 28 or 30 per cent" (Reports of Insp. of Fact.,
Dec. 1863, pp. 139, 142), whereas all this might be avoided through the
use of better machinery. So much tow fell by the wayside that the factory
inspector reports: "I have been informed with regard to some of the
scutch mills in Ireland, that the waste made at them has often been used

by the scutchers to burn on their fires at home, and yet it is very valuable"
(p. 140 of the above report). We shall speak of cotton waste later, when
we deal with the price fluctuations of raw materials.
  The wool industry was shrewder than the flax manufacturers. "It was
once the common practice to decry the preparation of waste and woollen
rags for re-manufacture, but the prejudice has entirely subsided as regards
the shoddy trade, which has become an important branch of the woollen
trade of Yorkshire, and doubtless the cotton waste trade will be recognised
in the same manner as supplying an admitted want. Thirty years since,
woollen rags, i.e., pieces of cloth, old clothes, etc., of nothing but wool,
would average about £4 4s. per ton in price: within the last few years they
have become worth £44 per ton, and the demand for them has so increased
that means have been found for utilising the rags of fabrics of cotton and
wool mixed by destroying the cotton and leaving the wool intact, and
now thousands of operatives are engaged in the manufacture of shoddy,
from which the consumer has greatly benefited in being able to purchase
cloth of a fair and average quality at a very moderate price." (Reports of
Insp. of Fact., Oct. 1863, p. 107.) By the end of 1862 the rejuvenated
shoddy made up as much as one-third of the entire consumption of wool
in English industry. (Reports of Insp. of Fact., October 1862, p. 81.)
The "big benefit" for the "consumer" is that his shoddy clothes wear
out in just one-third of the previous time and turn threadbare in one-sixth
of this time.
  The English silk industry moved along the same downward path. The
consumption of genuine raw silk decreased somewhat between 1839 and
1862, while that of silk waste doubled. Improved machinery helped to
manufacture a silk useful for many purposes from this otherwise rather
worthless stuff.
  The most striking example of utilising waste is furnished by the chemical industry. It utilises not only its own waste, for which it finds new uses,
but also that of many other industries. For instance, it converts the formerly almost useless gas-tar into aniline dyes, alizarin, and, more recently,
even into drugs.
  This economy of the excretions of production through their re-employment is to be distinguished from economy through the prevention of
waste, that is to say, the reduction of excretions of production to a minimum, and the immediate utilisation to a maximum of all raw and auxiliary
materials required in production.

Reduction of waste depends in part on the quality of the machinery in use. Economy in oil, soap, etc., depends on how well the mechanical parts are machined and polished. This refers to the auxiliary materials. In part, however, and this is most important, it depends on the quality of the employed machines and tools whether a larger or smaller portion of the raw material is turned into waste in the production process. Finally, this depends on the quality of the raw material itself. This, in turn, depends partly on the development of the extractive industry and agriculture which produce the raw material (strictly speaking on the progress of civilisation), and partly on the improvement of processes through which raw materials pass before they enter into manufacture.

Karl Marx, *Capital,* vol. 3, pp. 101-103.

F. In Altering Nature Man Produces Unforeseen and Harmful Consequences

Animals, as already indicated, change external nature by their activities just as man does, even if not to the same extent, and these changes made by them in their environment, as we have seen, in turn react upon and change their originators. For in nature nothing takes place in isolation. Everything affects every other thing and *vice-versa,* and it is mostly because this all-sided motion and interaction is forgotten that our natural scientists are prevented from clearly seeing the simplest things. We have seen how goats have prevented the regeneration of forests in Greece; on St. Helena, goats and pigs brought by the first navigators to arrive there succeeded in exterminating almost completely the old vegetation of the island, and so prepared the ground for the spreading of plants brought by later sailors and colonists. But if animals exert a lasting effect on their environment, it happens unintentionally, and as far as the animals themselves are concerned, it is an accident. The further men become removed from animals, however, the more their effect on nature assumes the character of premeditated, planned action directed toward definite ends known in advance. The animal destroys the vegetation of a locality without realizing what it is doing. Man destroys it in order to sow field crops on the soil thus released, or to plant trees or vines which he knows will yield many times the amount sown. He transfers useful plants and domestic animals from one country to another and thus changes the flora and fauna of whole continents. More than this. Through artificial breeding, both plants and animals are so changed by the hand of man that they become unrecognizable. The wild plants from which our grain varieties originated are still being sought in vain. The question of the wild animal from which our dogs are descended, the dogs themselves being so different from one another, or our equally numerous breeds of horses, is still under dispute.

It goes without saying, of course, that we have no intention of disputing the ability of animals to act in a planned and premeditated fashion. On the contrary, a planned mode of action exists in embryo wherever protoplasm, living protein, exists and reacts, i.e., carries out definite, even if extremely simple, movements as a result of definite external stimuli. Such reaction takes place even where there is as yet no cell at all, let alone a nerve cell. The manner in which insectivorous plants capture their prey appears likewise in a certain respect as a planned action, although performed quite unconsciously. In animals the capacity for conscious, planned action develops side by side with the development of the nervous system and among mammals it attains quite a high level. While fox-hunting in England, one can observe every day how unerringly the fox knows how to make use of its excellent knowledge of the locality in order to escape from its pursuers, and how well it knows and turns to account all favourable features of the ground that cause the scent to be lost. Among our domestic animals, more highly developed thanks to association with man, every day one can note acts of cunning on exactly the same level as those of children. For, just as the developmental history of the human embryo in the mother's womb is only an abbreviated repetition of the history, extending over millions of years, of the bodily evolution of our animal ancestors, beginning from the worm, so the mental development of the human child is only a still more abbreviated repetition of the intellectual development of these same ancestors, at least of the later ones. But all the planned action of all animals has never resulted in impressing the stamp of their will upon the world. It took man to do that.

In short, the animal merely *uses* external nature, and brings about changes in it simply by his presence; man by his changes makes it serve his ends, *masters* it. This is the final, essential distinction between man and other animals, and once again it is labour that brings about this distinction.

Let us not, however, flatter ourselves overmuch on account of our human conquests over nature. For each such conquest takes its revenge on us. Each of them, it is true, has in the first place the consequences on which we counted, but in the second and third places it has quite different,

unforeseen effects which only too often cancel out the first. The people
who, in Mesopotamia, Greece, Asia Minor, and elsewhere, destroyed the
forests to obtain cultivable land, never dreamed that they were laying the
basis for the present devastated condition of those countries, by removing
along with the forests the collecting centres and reservoirs of moisture.
When, on the southern slopes of the mountains, the Italians of the Alps
used up the fir forests so carefully cherished on the northern slopes, they
had no inkling that by doing so they were cutting at the roots of the dairy
industry in their region; they had still less inkling that they were thereby
depriving their mountain springs of water for the greater part of the year,
making it possible for these to pour still more furious flood torrents on
the plains during the rainy season. Those who spread the potato in Europe
were not aware that with these farinaceous tubers they were at the same
time spreading the disease of scrofula. Thus at every step we are reminded
that we by no means rule over nature like a conqueror over a foreign
people, like someone standing outside nature—but that we, with flesh,
blood, and brain, belong to nature, and exist in its midst, and that all our
mastery of it consists in the fact that we have the advantage over all other
creatures of being able to know and correctly apply its laws.

And, in fact, with every day that passes we are learning to understand
these laws more correctly, and getting to know both the more immediate
and the more remote consequences of our interference with the traditional
course of nature. In particular, after the mighty advances of natural science
in the present century, we are more and more placed in a position where
we can get to know, and hence to control, even the more remote natural
consequences at least of our most ordinary productive activities. But the
more this happens, the more will men once more not only feel, but also
know, themselves to be one with nature, and thus the more impossible
will become the senseless and anti-natural idea of a contradiction between
mind and matter, man and nature, soul and body, such as arose in Europe
after the decline of classic antiquity and which obtained its highest elaboration in Christianity.

But if it has already required the labour of thousands of years for us
to learn to some extent to calculate the more remote *natural* consequences
of our actions aiming at production, it has been still more difficult in regard to the more remote *social* consequences of these actions. We mentioned the potato and the resulting spread of scrofula. But what is scrofula
in comparison with the effect on the living conditions of the masses of
the people in whole countries that resulted from the workers being reduced

to a potato diet, or in comparison with the famine which overtook Ireland
in 1847 in consequence of the potato blight, and which consigned to the
grave a million Irishmen, nourished on potatoes and almost exclusively on
potatoes, and forced the emigration overseas of two million more? When
the Arabs learned to distil alcohol, it never entered their heads that by so
doing they were creating one of the chief weapons for the annihilation of
the aborigines of the then still undiscovered American continent. And
when afterwards Columbus discovered America, he did not know that by
doing so he was giving a new lease of life to slavery, which in Europe had
long ago been done away with, and laying the basis for the Negro slave
traffic. The men who in the seventeenth and eighteenth centuries laboured
to create the steam-engine had no idea that they were preparing the instrument which more than any other was to revolutionize social conditions
throughout the world. Especially in Europe, by concentrating wealth in
the hands of a minority, the huge majority being rendered propertyless,
this instrument was destined at first to give social and political domination
to the bourgeoisie, and then, however, to give rise to a class struggle between bourgeoisie and proletariat, which can end only in the overthrow
of the bourgeoisie and the abolition of all class antagonisms.—But in this
sphere too, by long and often cruel experience and by collecting and
analyzing the historical material, we are gradually learning to get a clear
view of the indirect, more remote social effects of our productive activity,
and so the possibility is afforded us of mastering and regulating these
effects as well.

To carry out this regulation requires something more than mere knowledge. It requires a complete revolution in our hitherto existing mode of
production and with it of our whole contemporary social order.

All hitherto existing modes of production have aimed merely at achieving the most immediately and directly useful effect of labour. The further
consequences, which only appear later on and become effective through
gradual repetition and accumulation, were totally neglected. The original
communal ownership of land corresponded, on the one hand, to a level
of development of human beings in which their horizon was restricted in
general to what lay immediately at hand, and presupposed, on the other
hand, a certain surplus of available land, allowing a certain latitude for
correcting any possible bad results of this primitive type of economy.
When this surplus land was exhausted, communal ownership also declined.
All higher forms of production, however, led to the division of the population into different classes and thereby to the antagonism of ruling and

oppressed classes. But thanks to this the interest of the ruling class became the driving factor of production, in so far as the latter was not
restricted to the barest means of subsistence of the oppressed people.
This has been carried through most completely in the capitalist mode of
production prevailing today in Western Europe. The individual capitalists,
who dominate production and exchange, are able to concern themselves
only with the most immediate useful effect of their actions. Indeed, even
this useful effect—in as much as it is a question of the usefulness of the
article that is produced or exchanged—retreats right into the background,
and the sole incentive becomes the profit to be gained on selling.

The social science of the bourgeoisie, classical political economy, is
predominantly occupied with only the directly intended social effects of
human actions aimed at production and exchange. This fully corresponds
to the social organization of which it is the theoretical expression. As
individual capitalists are engaged in production and exchange for the sake
of the immediate profit, only the nearest, most immediate results can be
taken into account in the first place. When an individual manufacturer or
merchant sells a manufactured or purchased commodity with the usual
profit he is satisfied, and he does not care what becomes afterwards of
the commodity and its purchasers. The same thing applies to the natural
effects of the same actions. What did it matter to the Spanish planters in
Cuba, who burned down forests on the slopes of the mountains and obtained from the ashes sufficient fertilizer for *one* generation of very highly
profitable coffee trees, what did it matter to them that the heavy tropical
rainfall afterwards washed away the now unprotected upper stratum of
the soil, leaving behind only bare rock? In relation to nature, as to society,
the present mode of production is predominantly concerned only about
the first, the most tangible result; and then surprise is expressed that the
more remote effects of actions directed to this end turn out to be of quite
a different, mainly even of quite an opposite, character; that the harmony
of demand and supply becomes transformed into its polar opposite as
shown by the course of each ten years' industrial cycle, and of which
Germany, too, has experienced a little prelude in the "crash"; that private
property based on individual labor necessarily develops into the property-
lessness of the workers, while all wealth becomes more and more concentrated in the hands of non-workers. . . .

Frederick Engels, *Dialectics of Nature,*
pp. 239-246.

G. Capitalism Wastes and Exhausts the Soil

Here, in small-scale agriculture, the price of land, a form and result of private landownership, appears as a barrier to production itself. In large-scale agriculture, and large estates operating on a capitalist basis, ownership likewise acts as a barrier, because it limits the tenant farmer in his productive investment of capital which in the final analysis benefits not him, but the landlord. In both forms, exploitation and squandering of the vitality of the soil (apart from making exploitation dependent upon the accidental and unequal circumstances of individual produce rather than the attained level of social development) takes the place of conscious rational cultivation of the soil as eternal communal property, an inalienable condition for the existence and reproduction of a chain of successive generations of the human race.

Karl Marx, *Capital*, vol. 3, p. 812.

Small landed property presupposes that the overwhelming majority of the population is rural, and that not social, but isolated labour predominates; and that, therefore, under such conditions wealth and development of reproduction, both of its material and spiritual prerequisites, are out of the question, and thereby also the prerequisites for rational cultivation. On the other hand, large landed property reduces the agricultural population to a constantly falling minimum, and confronts it with a constantly growing industrial population crowded together in large cities. It thereby creates conditions which cause an irreparable break in the coherence of social interchange prescribed by the natural laws of life. As a result, the vitality of the soil is squandered, and this prodigality is carried by commerce far beyond the borders of a particular state (Liebig).[1]

While small landed property creates a class of barbarians standing halfway outside of society, a class combining all the crudeness of primitive forms of society with the anguish and misery of civilised countries, large landed property undermines labour-power in the last region, where its prime energy seeks refuge and stores up its strength as a reserve fund for the regeneration of the vital force of nations—on the land itself. Large-scale industry and large-scale mechanised agriculture work together. If originally distinguished by the fact that the former lays waste and destroys principally

---

1. Liebig, *Die Chemie in ihrer Anwendung auf Agricultur und Physiologie*, Brunschweig, 1862—Ed.

labour-power, hence the natural force of human beings, whereas the latter more directly exhausts the natural vitality of the soil, they join hands in the further course of development in that the industrial system in the country-side also enervates the labourers, and industry and commerce on their part supply agriculture with the means for exhausting the soil.

Karl Marx, *Capital,* vol. 3, p. 813.

H. Deforestation Under Capitalism

The long production time (which comprises a relatively small period of working time) and the great length of the periods of turnover entailed make forestry an industry of little attraction to private and therefore capitalist enterprise, the latter being essentially private even if the associated capitalist takes the place of the individual capitalist. The development of culture and of industry in general has ever evinced itself in such energetic destruction of forests that everything done by it conversely for their preservation and restoration appears infinitesimal.

Karl Marx, *Capital,* vol. 2, p. 244.

I. Capitalism Ruins the Worker's Health and the Soil's Fertility

But in its blind unrestrainable passion, its were-wolf hunger for surplus-labour, capital oversteps not only the moral, but even the merely physical maximum bounds of the working-day. It usurps the time for growth, development, and healthy maintenance of the body. It steals the time required for the consumption of fresh air and sunlight. It higgles over a meal-time, incorporating it where possible with the process of production itself, so that food is given to the labourer as to a mere means of production, as coal is supplied to the boiler, grease and oil to the machinery. It reduces the sound sleep needed for the restoration, reparation, refreshment of the bodily powers to just so many hours of torpor as the revival of an organism, absolutely exhausted, renders essential. It is not the normal maintenance of the labour-power which is to determine the limits of the working-day; it is the greatest possible daily expenditure of labour-power, no matter how diseased, compulsory, and painful it may be, which is to determine the limits of the labourers' period of repose. Capital cares nothing for the length of life of labour-power. All that concerns it is simply and solely the maximum of labour-power, that can be rendered fluent in

a working-day. It attains this end by shortening the extent of the labourer's life, as a greedy farmer snatches increased produce from the soil by
robbing it of its fertility.

Karl Marx, Capital, vol. 1, pp. 264-265.

J. The Weeded Land, the Human Weeds

The cleanly weeded land, and the uncleanly human weeds, of Lincolnshire, are pole and counterpole of capitalistic production.

Karl Marx, Capital, vol. 1, p. 696.

IX. Capitalist Pollution of People in Places of Work and Dwellings

*In nineteenth-century England, many investigators, from the factory
inspectors of the government to popular writers like Charles Dickens, exposed the dreadful conditions of human life in the industrial factory and
in the slums of the working poor. Marx and Engels observed, studied, and
described such conditions too, but unlike their contemporaries they had a
general theory to explain and eradicate such conditions at the roots. They
did not yield to any contemporary in their capacity to confront and etch
in acidulous language the intolerable dehumanization of the capitalist
system. In contrast to their contemporaries, Marx and Engels perceived
the contradiction between the potential liberating power of industry, with
its machinery and science, and the degradation of laborers whose collective
energies and intelligence have raised up a system of wealth that serves only
to crush them under its burden. Although through the struggles of their
trade unions and political organizations the workers have made many
gains in the improvement of their wretched working and living conditions,
much yet remains to be done before such conditions are removed entirely.*

*As we read these graphic accounts by Marx and Engels, we wince and
wonder why things then so horrible were so endured. Yet, we should be
reminded that in the urban and rural slums of the "advanced" capitalism
of the United States, more than a century later, millions of people are still
condemned to live in squalor and poverty, to ache with hunger and waste
in illness, to have no clothes but rags, and to know no homes but shacks
and heatless wrecks overrun by rats and infested with roaches. Millions*

still work in places of duress, disease, and danger. Every year, noise, dust, dirt, fumes, gases, toxic chemicals, heat, radiation, and unsafe machinery pollute, make ill, injure, and kill more than half of the 80 million workers in our country. Here, in the bodies of workers, is where the corrupt ecology of capitalism hits home. Here is where Marx and Engels saw it at its worst. And here is where it must first be expunged, by the workers themselves.
                                                                                    —Ed.

A. As Man Masters Nature, He Is Enslaved to Others

The so-called revolutions of 1848 were but poor incidents, small fractures and fissures in the dry crust of European society. However, they denounced the abyss. Beneath the apparently solid surface, they betrayed oceans of liquid matter, only needing expansion to rend into fragments continents of hard rock. Noisily and confusedly they proclaimed the emancipation of the proletarian, i.e., the secret of the nineteenth century, and of the revolution of that century. The social revolution, it is true, was no novelty invented in 1848. Steam, electricity and the self-acting mule were revolutions of a rather more dangerous character than even Citizens Barbès, Raspail, and Blanqui! But, although the atmosphere in which we live weighs upon everyone with a twenty-thousand-pound force, do you feel it? No more than European society before 1848 felt the revolutionary atmosphere enveloping it and pressing it from all sides. There is one great fact characteristic of this our nineteenth century, a fact which no party dares deny. On the one hand there have started into life industrial and scientific forces which no epoch of the former human history had ever suspected. On the other hand there exist symptoms of decay, far surpassing the horrors recorded of the latter times of the Roman Empire. In our days, everything seems pregnant with its contrary. Machinery, gifted with the wonderful power of shortening and fructifying human labour, we behold starving and overworking it. The newfangled sources of wealth, by some strange, weird spell, are turned into sources of want. The victories of art seem bought by the loss of character. At the same pace that mankind masters nature, man seems to become enslaved to other men or to his own infamy. Even the pure life of science seems unable to shine but on the dark background of ignorance. All our invention and progress seem to result in endowing material forces with intellectual life, and in stultifying human life into a material force. This antagonism between modern industry and science, on the one hand, and modern misery and dissolution, on the other hand; this antagonism between the productive forces and the social relations of our epoch is a fact, palpable, overwhelming, and not to be controverted. Some may wail over it; others may wish to get rid of modern arts, in order to get rid of modern conflicts. Or they may imagine that so signal a progress in industry wants to be completed by as signal a regress in politics. For our part, we do not mistake the shape of the shrewd spirit that continues to mark all these contradictions. We know that if the newfangled forces of society are to work satisfactorily, they need only be mastered by newfangled men—and such are the working men. They are as much the invention of modern time as machinery itself. In the signs that bewilder the middle class, the aristocracy, and the poor prophets of regression, we recognise our old friend Robin Goodfellow, the old mole that can work in the earth so fast, that worthy pioneer—the revolution. The English working men are the firstborn sons of modern industry. Certainly, then, they will not be the last to aid the social revolution produced by that industry—a revolution which means the emancipation of their class all over the world, which is as universal as capital-rule and wage slavery. I know the heroic struggles the English working class has gone through since the middle of the last century; struggles not the less glorious because they are shrouded in obscurity and burked by middle-class historians. To take vengeance for the misdeeds of the ruling class there existed in the Middle Ages in Germany a secret tribunal called the *Vehmgericht*. If a red cross was seen marked on a house, people knew that its owner was doomed by the *Vehm*. All the houses of Europe are now marked by the mysterious red cross. History is the judge; its executioner, the proletarian.

> Karl Marx, Speech of April 14, 1856,
> in Karl Marx and Frederick Engels,
> *Selected Correspondence 1846-1895,*
> pp. 90-91.

B. Exploitative and Unhealthful Conditions in the Industrial Factory and Workers' Dwelling

Although then, technically speaking, the old system of division of labour is thrown overboard by machinery, it hangs on in the factory, as a traditional habit handed down from Manufacture, and is afterwards systematically re-moulded and established in a more hideous form by capital as a means of exploiting labour-power. The life-long specialty of handling one and the same tool, now becomes the life-long speciality of serving one and the same machine. Machinery is put to a wrong use, with the object of transforming the workman, from his very childhood, into a part of a detail-machine.[1] In this way, not only are the expenses of his reproduction considerably lessened, but at the same time his helpless dependence upon the factory as a whole, and therefore upon the capitalist, is rendered complete. Here as everywhere else, we must distinguish between the increased productiveness due to the development of the social process of production, and that due to the capitalist exploitation of that process. In handicrafts and manufacture, the workman makes use of a tool, in the factory, the machine makes use of him. There the movements of the instrument of labour proceed from him, here it is the movements of the machine that he must follow. In manufacture the workmen are parts of a living mechanism. In the factory we have a lifeless mechanism independent of the workman, who becomes its mere living appendage. "The miserable routine of endless drudgery and toil in which the same mechanical process is gone through over and over again, is like the labour of Sisyphus. The burden of labour, like the rock, keeps ever falling back on the worn-out labourer."[2] At the same time that factory work exhausts the nervous to the uttermost, it does away with the many-sided play of the muscles, and confiscates every atom of freedom, both in bodily and intellectual activity.[3] The lightening of the labour, even, becomes a sort of torture, since the machine does not free the labourer from work, but deprives the work of all interest. Every kind of capitalist production, in so far as it is not only a labour-process, but also a process of creating surplus-value, has this in common, that it is not the workman

---

1. So much then for Proudhon's wonderful idea: he "construes" machinery not as a synthesis of instruments of labour, but as a synthesis of detail operations for the benefit of the labourer himself.

2. F. Engels, 1. c., p. 217. Even an ordinary and optimistic Free-trader, like Mr. Molinari goes so far as to say, "Un homme s'use plus vite en surveillant, quinze heures par jour, l'évolution uniforme d'un mécanisme, qu'en exerçant, dans le même espace de temps, sa force physique. Ce travail de surveillance qui servirait peut-être d'utile gymnastique à l'intelligence, s'il n'était pas trop prolongé, détruit à la longue, par son excès et l'intelligence, et le corps même." (G. de Molinari: "Etudes Economiques." Paris, 1846.)

3. F. Engels, 1. c., p. 216.

that employs the instruments of labour, but the instruments of labour that employ the workman. But it is only in the factory system that this inversion for the first time acquires technical and palpable reality. By means of its conversion into an automaton, the instrument of labour confronts the labourer, during the labour-process, in the shape of capital, of dead labour, that dominates, and pumps dry, living labour-power. The separation of the intellectual powers of production from the manual labour, and the conversion of those powers into the might of capital over labour, is, as we have already shown, finally completed by modern industry erected on the foundation of machinery. The special skill of each individual insignificant factory operative vanishes as an infinitesimal quantity before the science, the gigantic physical forces, and the mass of labour that are embodied in the factory mechanism and, together with that mechanism, constitute the power of the "master." This "master," therefore, in whose brain the machinery and his monopoly of it are inseparably united, whenever he falls out with his "hands," contemptuously tells them: "The factory operatives should keep in wholesome remembrance the fact that theirs is really a low species of skilled labour; and that there is none which is more easily acquired, or of its quality more amply remunerated, or which by a short training of the least expert can be more quickly, as well as abundantly, acquired. . . . The master's machinery really plays a far more important part in the business of production than the labour and the skill of the operative, which six months' education can teach, and a common labourer can learn."[4] The technical subordination of the workman to the uniform motion of the instruments of labour, and the peculiar composition of the body of workpeople, consisting as it does of individuals of both sexes and of all ages, give rise to a barrack discipline, which is elaborated into a complete system in the factory, and which fully develops the before mentioned labour of overlooking, thereby dividing the workpeople into operatives and overlookers, into private soldiers and sergeants of an industrial army. "The main difficulty [in the automatic factory] . . . lay . . . above all in training human beings to renounce their desultory habits of work, and to identify themselves with the unvarying regularity of the complex automaton. To devise

---

4. "The Master Spinners' and Manufacturers' Defence Fund. Report of the Committee." Manchester, 1854, p. 17. We shall see hereafter, that the "master" can sing quite another song, when he is threatened with the loss of his "living" automaton.

and administer a successful code of factory discipline, suited to the necessities of factory diligence, was the Herculean enterprise, the noble
achievement of Arkwright! Even at the present day, when the system
is perfectly organised and its labour lightened to the utmost, it is found
nearly impossible to convert persons past the age of puberty, into useful
factory hands."[5] The factory code in which capital formulates, like a
private legislator, and at his own good will, his autocracy over his workpeople, unaccompanied by that division of responsibility, in other matters so much approved of by the bourgeoisie, and unaccompanied by
the still more approved representative system, this code is but the capitalistic caricature of that social regulation of the labour-process which
becomes requisite in co-operation on a great scale, and in the employment in common, of instruments of labour and especially of machinery.
The place of the slave-driver's lash is taken by the overlooker's book of
penalties. All punishments naturally resolve themselves into fines and
deductions from wages, and the law-giving talent of the factory Lycurgus
so arranges matters, that a violation of his laws is, if possible, more profitable to him than the keeping of them.[6]

---

5. Ure, 1. c., p. 15. Whoever knows the life history of Arkwright, will never
dub this barber-genius "noble." Of all the great inventors of the 18th century,
he was incontestably the greatest thiever of other people's inventions and the
meanest fellow.

6. "The slavery in which the bourgeoisie has bound the proletariat, comes nowhere more plainly into daylight than in the factory system. In it all freedom
comes to an end both at law and in fact. The workman must be in the factory at
half past five. If he come a few minutes late, he is punished; if he come 10 minutes
late, he is not allowed to enter until after breakfast, and thus loses a quarter of a
day's wage. He must eat, drink and sleep at word of command . . . The despotic
bell calls him from his bed, calls him from breakfast and dinner. And how does he
fare in the mill? There the master is the absolute law-giver. He makes what regulations he pleases; he alters and makes additions to his code at pleasure; and if he
insert the veriest nonsense, the courts say to the workman: Since you have entered
into this contract voluntarily, you must now carry it out . . . These workmen are
condemned to live, from their ninth year till their death, under this mental and
bodily torture." (F. Engels, 1. c., p. 217, sq.) What, "the courts say," I will illustrate
by two examples. One occurs at Sheffield at the end of 1866. In that town a workman had engaged himself for 2 years in a steelworks. In consequence of a quarrel
with his employer he left the works, and declared that under no circumstances
would he work for that master any more. He was prosecuted for breach of contract,
and condemned to two months' imprisonment. (If the master break the contract,

We shall here merely allude to the material conditions under which factory labour is carried on. Every organ of sense is injured in an equal degree by artificial elevation of the temperature, by the dust-laden atmosphere, by the deafening noise, not to mention danger to life and limb among the thickly crowded machinery, which, with the regularity of the seasons, issues its list of the killed and wounded in the industrial battle.[7] Economy of the social means of production, matured and forced as in a hothouse by the factory system, is turned, in the hands of capital, into systematic robbery of what is necessary for the life of the workman while he is at work, robbery of space, light, air, and of protection to his person against the dangerous and unwholesome accompaniments of the productive process, not to mention the robbery of appliances for the comfort of the workman.[8] Is Fourier wrong when he calls factories "tempered bagnos"?[9]

Karl Marx, *Capital,* vol. 1, pp. 422-427.

Of these facts, Dr. Simon in his General Health Report says: "That cases are innumerable in which defective diet is the cause or the aggra-

---

he can be proceeded against only in a civil action, and risks nothing but money damages.) After the workman has served his two months, the master invites him to return to the works, pursuant to the contract. Workman says: No, he has already been punished for the breach. The master prosecutes again, the court condemns again, although one of the judges, Mr. Shee, publicly denounces this as a legal monstrosity, by which a man can periodically, as long as he lives, be punished over and over again for the same offence or crime. This judgment was given not by the "Great Unpaid," the provincial Dogberries, but by one of the highest courts of justice in London.—[*Added in the 4th German edition.*—This has now been done away with. With few exceptions, *e.g.,* when public gas-works are involved, the worker in England is now put on an equal footing with the employer in case of breach of contract and can be sued only civilly.—F. E.] The second case occurs in Wiltshire at the end of November 1863. About 30 power-loom weavers, in the employment of one Harrup, a cloth manufacturer at Leower's Mill, Westbury Leigh, struck work because master Harrup indulged in the agreeable habit of making deductions from their wages for being late in the morning; 6d. for 2 minutes; 1s. for 3 minutes, and 1s. 6d. for ten minutes. This is at the rate of 9s. per hour, and £4 10s. 0d. per diem; while the wages of the weavers on the average of a year, never exceeded 10s. to 12s. weekly. Harrup also appointed a boy to announce the starting time by a whistle, which he often did before six o'clock in the morning: and if the hands were not all there at the moment the whistle ceased, the doors were closed, and those hands who were outside were fined: and as there was no clock on the premises, the unfortunate hands were at the mercy of the young Harrup-inspired vator of disease, can be affirmed by any one who is conversant with poor law medical practice, or with the wards and out-patient rooms of hospitals. . . . Yet in this point of view, there is, in my opinion, a very important sanitary context to be added. It must be remembered that privation of

---

time-keeper. The hands on strike, mothers of families as well as girls, offered to resume work if the time-keeper were replaced by a clock, and a more reasonable scale of fines were introduced. Harrup summoned 19 women and girls before the magistrates for breach of contract. To the utter indignation of all present, they were each mulcted in a fine of 6d. and 2s. 6d. for costs. Harrup was followed from the court by a crowd of people who hissed him.—A favourite operation with manufacturers is to punish the workpeople by deductions made from their wages on account of faults in the material worked on. This method gave rise in 1886 to a general strike in the English pottery districts. The reports of the Ch. Empl. Com. (1863-1866), give cases where the worker not only receives no wages, but becomes, by means of his labour, and of the penal regulations, the debtor to boot, of his worthy master. The late cotton crisis also furnished edifying examples of the sagacity shown by the factory autocrats in making deductions from wages. Mr. R. Baker, the Inspector of Factories, says, "I have myself had lately to direct prosecutions against one cotton mill occupier for having in these pinching and painful times deducted 10d. a piece from some of the young workers employed by him, for the surgeon's certificate (for which he himself had only paid 6d.), when only allowed by the law to deduct 3d., and by custom nothing at all . . . And I have been informed of another, who in order to keep without the law, but to attain the same object, charges the poor children who work for him a shilling each, as a fee for learning them the art and mystery of cotton spinning, so soon as they are declared by the surgeon fit and proper persons for that occupation. There may therefore be undercurrent causes for such extraordinary exhibitions as strikes, not only wherever they arise, but particularly at such times as the present, which without explanation, render them inexplicable to the public understanding." He alludes here to a strike of power-loom weavers at Darwen, June, 1863. ("Reports of Insp. of Fact. for 30 April, 1863," pp. 50-51.) The reports always go beyond their official dates.

7. The protection afforded by the Factory Acts against dangerous machinery has had a beneficial effect. "But . . . there are other sources of accident which did not exist twenty years since; one especially, viz., the increased speed of the machinery. Wheels, rollers, spindles and shuttles are now propelled at increased and increasing rates; fingers must be quicker and defter in their movements to take up the broken thread, for, if placed with hesitation or carelessness, they are sacrificed. . . . A large number of accidents are caused by the eagerness of the workpeople to get through their work expeditiously. It must be remembered that it is of the highest importance to manufacturers that their machinery should be in motion, *i.e.,* producing yarns and goods. Every minute's stoppage is not only a loss of power, but of production and the workpeople are urged by the overlookers, who are interested in the quantity of work turned off, to keep the machinery in motion;

food is very reluctantly borne, and that as a rule great poorness of diet will only come when other privations have preceded it. Long before insufficiency of diet is a matter of hygienic concern, long before the physiologist would think of counting the grains of nitrogen and carbon which

---

and it is no less important to those of the operatives who are paid by the weight or piece, that the machines should be kept in motion. Consequently, although it is strictly forbidden in many, nay in most factories, that machinery should be cleaned while in motion, it is nevertheless the constant practice in most, if not in all, that the workpeople do, unreproved, pick out waste, wipe rollers and wheels, &c., while their frames are in motion. Thus from this cause only, 906 accidents have occurred during the six months.... Although a great deal of cleaning is constantly going on day by day, yet Saturday is generally the day set apart for the thorough cleaning of the machinery, and a great deal of this is done while the machinery is in motion." Since cleaning is not paid for, the workpeople seek to get done with it as speedily as possible. Hence "the number of accidents which occur on Fridays, and especially on Saturdays, is much larger than on any other day. On the former day the excess is nearly 12 per cent. over the average number of the four first days of the week, and on the latter day the excess is 25 per cent. over the average of the preceding five days; or, if the number of working hours on Saturday being taken into account—7½ hours on Saturday as compared with 10½ on the other days—there is an excess of 65 per cent. on Saturdays over the average of the other five days." ("Rep. of Insp. of Fact., 31st Oct., 1866," pp. 9, 15, 16, 17.)

8. In Part I. of Book III. I shall give an account of a recent campaign by the English manufacturers against the Clauses in the Factory Acts that protect the "hands" against dangerous machinery. For the present, let this one quotation from the official report of Leonard Horner suffice: "I have heard some mill-owners speak with inexcusable levity of some of the accidents; such, for instance, as the loss of a finger being a trifling matter. A working man's living and prospects depend so much upon his fingers, that any loss of them is a very serious matter to him. When I have heard such inconsiderate remarks made, I have usually put this question: Suppose you were in want of an additional workman, and two were to apply, both equally well qualified in other respects, but one had lost a thumb or a forefinger, which would you engage? There never was a hesitation as to the answer." ... The manufacturers have "mistaken prejudices against what they have heard represented as a pseudo-philanthropic legislation." ("Rep. of Insp. of Fact., 31st Oct., 1855.") These manufacturers are clever folk, and not without reason were they enthusiastic for the slave-holders' rebellion.

9. In those factories that have been longest subject to the Factory Acts, with their compulsory limitation of the hours of labour, and other regulations, many of the older abuses have vanished. The very improvement of the machinery demands to a certain extent "improved construction of the building," and this is an advantage to the workpeople. (See "Rep. of Insp. of Fact. for 31st Oct., 1863," p. 109.)

intervene between life and starvation, the household will have been utterly destitute of material comfort; clothing and fuel will have been even scantier than food—against inclemencies of weather there will have been no adequate protection—dwelling space will have been stinted to the degree in which over-crowding produces or increases disease; of household utensils and furniture there will have been scarcely any—even cleanliness will have been found costly or difficult, and if there still be self-respectful endeavours to maintain it, every such endeavour will represent additional pangs of hunger. The home, too, will be where shelter can be cheapest bought; in quarters where commonly there is least fruit of sanitary supervision, least drainage, least scavenging, least suppression of public nuisances, least or worst water supply, and, if in town, least light and air. Such are the sanitary dangers to which poverty is almost certainly exposed, when it is poverty enough to imply scantiness of food. And while the sum of them is of terrible magnitude against life, the mere scantiness of food is in itself of very serious moment. . . . These are painful reflections, especially when it is remembered that the poverty to which they advert is not the deserved poverty of idleness. In all cases it is the poverty of working populations. Indeed, as regards the indoor operatives, the work which obtains the scanty pittance of food, is for the most part excessively prolonged. Yet evidently it is only in a qualified sense that the work can be deemed self-supporting. . . . And on a very large scale the nominal self-support can be only a circuit, longer or shorter, to pauperism."[1]

The intimate connexion between the pangs of hunger of the most industrious layers of the working-class, and the extravagant consumption, coarse or refined, of the rich, for which capitalist accumulation is the basis, reveals itself only when the economic laws are known. It is otherwise with the "housing of the poor." Every unprejudiced observer sees that the greater the centralisation of the means of production, the greater is the corresponding heaping together of the labourers, within a given space; that therefore the swifter capitalistic accumulation, the more miserable are the dwellings of the working-people. "Improvements" of towns, accompanying the increase of wealth, by the demolition of badly built quarters, the erection of palaces for banks, warehouses, &c., the widening of streets for business traffic, for the carriages of luxury,

---

1. "Public Health. Sixth Report, 1864," pp. 14, 15.

and for the introduction of tramways, &c., drive away the poor into even worse and more crowded hiding places. On the other hand, every one knows that the dearness of dwellings is in inverse ration to their excellence, and that the mines of misery are exploited by house speculators with more profit or less cost than ever were the mines of Potosi. The antagonistic character of capitalist accumulation, and therefore of the capitalistic relations of property generally,[2] is here so evident, that even the official English reports on this subject teem with heterodox onslaughts on "property and its rights." With the development of industry, with the accumulation of capital, with the growth and "improvement" of towns, the evil makes such progress that the mere fear of contagious diseases which do not spare even "respectability," brought into existence from 1847 to 1864 no less than 10 Acts of Parliament on sanitation, and that the frightened bourgeois in some towns, as Liverpool, Glasgow, &c., took strenuous measures through their muncipalities. Nevertheless Dr. Simon, in his report of 1865, says: "Speaking generally, it may be said that the evils are uncontrolled in England." By order of the Privy Council, in 1864, an inquiry was made into the conditions of the housing of the agricultural labourers, in 1865 of the poorer classes in the towns. The results of the admirable work of Dr. Julian Hunter are to be found in the seventh (1865) and eighth (1866) reports on "Public Health." To the agricultural labourers, I shall come later. On the condition of town dwellings, I quote, as preliminary, a general remark of Dr. Simon. "Although my official point of view," he says, "is one exclusively physical, common humanity requires that the other aspect of this evil should not be ignored. . . . In its higher degrees it [i.e., over-crowding] almost necessarily involves such negation of all delicacy, such unclean confusion of bodies and bodily functions, such exposure of animal and sexual nakedness, as is rather bestial than human. To be subject to these influences is a degradation which must become deeper and deeper for those on whom it continues to work. To children who are born under its curse, it must often be a very baptismal into infamy. And beyond all measure hopeless is the wish that persons thus circumstanced should

---

2. "In no particular have the rights of *persons* been so avowedly and shamefully sacrificed to the rights of *property* as in regard to the lodging of the labouring class. Every large town may be looked upon as a place of human sacrifice, a shrine where thousands pass yearly through the fire as offerings to the moloch of avarice." S. Laing, 1. c., p. 150.

ever in other respects aspire to that atmosphere of civilisation which has its essence in physical and moral cleanliness."[3]

London takes the first place in over-crowded habitations, absolutely unfit for human beings. "He feels clear," says Dr. Hunter, "on two points; first, that there are about 20 large colonies in London, of about 10,000 persons each, whose miserable condition exceeds almost anything he has seen elsewhere in England, and is almost entirely the result of their bad house accomodation; and second, that the crowded and dilapidated condition of the houses of these colonies is much worse than was the case 20 years ago."[4] "It is not too much to say that life in parts of London and Newcastle is infernal."[5]

Further, the better-off part of the working-class, together with the small shopkeepers and other elements of the lower middle class, falls in London more and more under the curse of these vile conditions of dwelling, in proportion as "improvements," and with them the demolition of old streets and houses, advance, as factories and the afflux of human beings grow in the metropolis, and finally as house rents rise with the ground-rents. "Rents have become so heavy that few labouring men can afford more than one room."[6] There is almost no house-property in London that is not overburdened with a number of middlemen. For the price of land in London is always very high in comparison with its yearly revenue, and therefore every buyer speculates on getting rid of it again at a jury price (the expropriation valuation fixed by jurymen), or on pocketing an extraordinary increase of value arising from the neighbourhood of some large establishment. As a consequence of this there is a regular trade in the purchase of "fag-ends of leases." "Gentlemen in this business

---

3. "Public Health, eighth report, 1866," p. 14, note.

4. 1. c., p. 89. With reference to the children in these colonies, Dr. Hunter says: "People are not now alive to tell us how children were brought up before this age of dense agglomerations of poor began, and he would be a rash prophet who should tell us what future behaviour is to be expected from the present growth of children, who, under circumstances probably never paralleled in this country, are now completing their education for future practice, as 'dangerous classes' by sitting up half the night with persons of every age, half naked, drunken, obscene, and quarrelsome." (1. c., p. 56.)

5. 1. c., p. 62.

6. "Report of the Officer of Health of St. Martins-in-the-Fields, 1865."

may be fairly expected to do as they do—get all they can from the tenants while they have them, and leave as little as they can for their successors."[7]

The rents are weekly, and these gentlemen run no risk. In consequence of the making of railroads in the City, "the spectacle has lately been seen in the East of London of a number of families wandering about some Saturday night with their scanty worldly goods on their backs, without any resting place but the workhouse."[8] The workhouses are already over-crowded, and the "improvements" already sanctioned by Parliament are only just begun. If labourers are driven away by the demolition of their old houses, they do not leave their old parish, or at most they settle down on its borders, as near as they can get to it. "They try, of course, to remain as near as possible to their workshops. The inhabitants do not go beyond the same or the next parish, parting their two-room tenements into single rooms, and crowding even those. . . . Even at an advanced rent, the people who are displaced will hardly be able to get an accomodation so good as the meagre one that they have left. . . . Half the workmen . . . of the Strand . . . walked two miles to their work."[9] This same Strand, a main thoroughfare which gives strangers an imposing idea of the wealth of London, may serve as an example of the packing together of human beings in that town. In one of its parishes, the Officer of Health reckoned 581 persons per acre, although half the width of the Thames was reckoned in. It will be self-understood that every sanitary measure, which, as has been the case hitherto in London, hunts the labourers from one quarter by demolishing uninhabitable houses, serves only to crowd them together yet more closely in another. "Either," says Dr. Hunter, "the whole proceeding will of necessity stop as an absurdity, or the public compassion (!) be effectually aroused to the obligation which may now be without exaggeration called national, of supplying cover to those who by reason of their having no capital, cannot provide it for themselves, though they can by periodical payments reward those who will provide it for them."[10] Admire this capitalistic justice! The owner of land, of houses, the businessman, when expropriated by "improvements" such as railroads, the building of new streets, &c., not only

7. Public Health, eighth report, 1866, p. 91.

8. l. c., p. 88.

9. l. c., p. 88.

10. l. c., p. 89.

receives full idemnity. He must, according to law, human and divine, be comforted for his enforced "abstinence" over and above this by a thumping profit. The labourer, with his wife and child and chattels, is thrown out into the street, and—if he crowds in too large numbers towards quarters of the town where the vestries insist on decency, he is prosecuted in the name of sanitation!

Except London, there was at the beginning of the 19th century no single town in England of 100,000 inhabitants. Only five had more than 50,000. Now there are 28 towns with more than 50,000 inhabitants. "The result of this change is not only that the class of town people is enormously increased, but the old closepacked little towns are now centres, built round on every side, open nowhere to air, and being no longer agreeable to the rich are abandoned by them for the pleasanter outskirts. The successors of these rich are occupying the larger houses at the rate of a family to each room [. . . and find accommodation for two or three lodgers . . .] and a population, for which the houses were not intended and quite unfit, has been created, whose surroundings are truly degrading to the adults and ruinous to the children."[11] The more rapidly capital accumulates in an industrial or commercial town, the more rapidly flows the stream of exploitable human material, the more miserable are the improvised dwellings of the labourers.

Newcastle-on-Tyne, as the centre of a coal and iron district of growing productiveness, takes the next place after London in the housing inferno. Not less than 34,000 persons live there in single rooms. Because of their absolute danger to the community, houses in great numbers have lately been destroyed by the authorities in Newcastle and Gateshead. The building of new houses progresses very slowly, business very quickly. The town was, therefore, in 1865, more full than ever. Scarcely a room was to let. Dr. Embleton, of the Newcastle Fever Hospital, says: "There can be little doubt that the great cause of the continuance and spread of the typhus has been the over-crowding of human beings, and the uncleanliness of their dwellings. The rooms, in which labourers in many cases live, are situated in confined and unwholesome yards or courts, and for space, light, air and cleanliness, are models of insufficiency and insalubrity, and a disgrace to any civilised community; in them men, women, and children lie at night huddled together; and as regards the

---

11. 1. c., pp. 55 and 56.

men, the night-shift succeed the day-shift, and the day-shift the night-shift in unbroken series for some time together, the beds having scarcely time to cool; the whole house badly supplied with water and worse with privies, dirty, unventilated, and pestiferous."[12] The price per week of such lodgings ranges from 8d to 3s. "The town of Newcastle-on-Tyne," says Dr. Hunter, "contains a sample of the finest tribe of our country-men, often sunk by external circumstances of house and street into an almost savage degradation."[13]

As a result of the ebbing and flowing of capital and labour, the state of the dwellings of an industrial town may to-day be bearable, to-morrow hideous. Or the aedileship of the town may have pulled itself together for the removal of the most shocking abuses. To-morrow, like a swarm of locusts, come crowding in masses of ragged Irishmen or decayed English agricultural labourers. They are stowed away in cellars and lofts, or the hitherto respectable labourer's dwelling is transformed into a lodging-house whose *personnel* changes as quickly as the billets in the 30 years' war. Example: Bradford (Yorkshire). There the municipal philistine was just busied with urban improvements. Besides, there were still in Brad-ford, in 1861, 1,751 uninhabited houses. But now comes that revival of trade which the mildly liberal Mr. Forster, the negro's friend, recently crowed over with so much grace. With the revival of trade came of course an overflow from the waves of the ever fluctuating "reserve army" or "relative surplus-population." The frightful cellar habitations and rooms registered in the list, which Dr. Hunter obtained from the agent of an Insurance Company, were for the most part inhabited by well-paid labourers. They declared that they would willingly pay for better dwell-ings if they were to be had. Meanwhile, they become degraded, they fall ill, one and all, whilst the mildly liberal Forster, M. P., sheds tears over the blessings of Free-trade, and the profits of the eminent men of Brad-ford who deal in worsted. In the Report of September, 1865, Dr. Bell, one of the poor law doctors of Bradford, ascribes the frightful mortality of fever-patients in his district to the nature of their dwellings. "In one small cellar measuring 1,500 cubic feet . . . there are ten persons. . . . Vincent Street, Green Aire Place, and the Leys include 223 houses having 1,450 inhabitants, 435 beds, and 36 privies. . . . The beds—and in that

12. 1. c., p. 149.

13. 1. c., p. 50.

term I include any roll of dirty old rags, or an armful of shavings—have an average of 3.3 persons to each, many have 5 and 6 persons to each, and some people, I am told, are absolutely without beds; they sleep in their ordinary clothes, on the bare boards—young men and women, married and unmarried, all together. I need scarcely add that many of these dwellings are dark, damp, dirty, stinking holes, utterly unfit for human habitations; they are the centres from which disease and death are distributed amongst those in better circumstances, who have allowed them thus to fester in our midst."[14]

Bristol takes the third place after London in the misery of its dwellings. "Bristol, where the blankest poverty and domestic misery abound in the wealthiest town of Europe."[15]

Karl Marx, *Capital,* vol. 1, pp. 656-663.

C. The Poor Health of Industrial Workers

That a class which lives under the conditions already sketched and is so ill-provided with the most necessary means of subsistence, cannot be healthy and can reach no advanced age, is self-evident. Let us review the circumstances once more with especial reference to the health of the workers. The centralisation of population in great cities exercises of itself an unfavourable influence; the atmosphere of London can never be so pure, so rich in oxygen, as the air of the country; two and a half million pairs of lungs, two hundred and fifty thousand fires, crowded upon an area three to four miles square, consume an enormous amount of oxygen, which is replaced with difficulty, because the method of building cities in itself impedes ventilation. The carbonic acid gas, engendered by respiration and fire, remains in the streets by reason of its specific gravity, and the chief air current passes over the roofs of the city. The lungs of the inhabitants fail to receive the due supply of oxygen, and the consequence is mental and physical lassitude and low vitality. For this reason, the dwellers in cities are far less exposed to acute, and especially to inflammatory, affections than rural populations, who live in a free, normal atmosphere; but they suffer the more from chronic affec-

---

14. 1. c., p. 114.

15. 1. c., p. 50.

tions. And if life in large cities is, in itself, injurious to health, how great must be the harmful influence of an abnormal atmosphere in the working-people's quarters, where, as we have seen, everything combines to poison the air. In the country, it may, perhaps, be comparatively innoxious to keep a dung-heap adjoining one's dwelling, because the air has free ingress from all sides; but in the midst of a large town, among closely built lanes and courts that shut out all movement of the atmosphere, the case is different. All putrefying vegetable and animal substances give off gases decidedly injurious to health, and if these gases have no free way of escape, they inevitably poison the atmosphere. The filth and stagnant pools of the working-people's quarters in the great cities have, therefore, the worst effect upon the public health, because they produce precisely those gases which engender disease; so, too, the exhalations from contaminated streams. But this is by no means all. The manner in which the great multitude of the poor is treated by society to-day is revolting. They are drawn into the large cities where they breathe a poorer atmosphere than in the country; they are relegated to districts which, by reason of the method of construction, are worse ventilated than any others; they are deprived of all means of cleanliness, of water itself, since pipes are laid only when paid for, and the rivers so polluted that they are useless for such purposes; they are obliged to throw all offal and garbage, all dirty water, often all disgusting drainage and excrement into the streets, being without other means of disposing of them; they are thus compelled to infect the region of their own dwellings. Nor is this enough. All conceivable evils are heaped upon the heads of the poor. If the population of great cities is too dense in general, it is they in particular who are packed into the least space. As though the vitiated atmosphere of the streets were not enough, they are penned in dozens into single rooms, so that the air which they breathe at night is enough in itself to stifle them. They are given damp dwellings, cellar dens that are not waterproof from below, or garrets that leak from above. Their houses are so built that the clammy air cannot escape. They are supplied bad, tattered, or rotten clothing, adulterated and indigestible food. They are exposed to the most exciting changes of mental condition, the most violent vibrations between hope and fear; they are hunted like game, and not permitted to attain peace of mind and quiet enjoyment of life. They are deprived of all enjoyments except that of sexual indulgence and drunkeness, are worked every day to the point of complete exhaustion of their mental and physical energies, and are thus constantly spurred on to the maddest excess in the only two enjoyments at their command. And if they surmount all this, they fall victims to want of work in a crisis when all the little is taken from them that had hitherto been vouchsafed them.

How is it possible, under such conditions, for the lower class to be healthy and long lived? What else can be expected than an excessive mortality, an unbroken series of epidemics, a progressive deterioration in the physique of the working population?

> Frederick Engels, *The Condition of*
> *the Working-Class in England in 1844,*
> pp. 96-98.

D. *Urban Housing Reform and the Perpetuation of Slums*

By "Haussmann" I mean the practice which has now become general of making breaches in the working-class quarters of our big towns, and particularly in those which are centrally situated, quite apart from whether this is done from considerations of public health and for beautifying the town, or owing to the demand for big centrally situated business premises, or owing to traffic requirements, such as the laying down of railways, streets, etc. No matter how different the reasons may be, the result is everywhere the same: the scandalous alleys and lanes disappear to the accompaniment of lavish self-praise from the bourgeoisie on account of this tremendous success, but they appear again immediately somewhere else and often in the immediate neighborhood.

In *The Condition of the Working Class in England* I gave a description of Manchester as it looked in 1843 and 1844. Since then the construction of railways through the center of the town, the laying out of new streets, and the erection of great public and private buildings have broken through, laid bare and improved some of the worst districts described in my book, others have been abolished altogether, but many of them are still, apart from the fact that official sanitary inspection has since become stricter, in the same state or in an even worse state of dilapidation than they were then. On the other hand, however, thanks to the enormous extension of the town, whose population has increased since then by more than half, districts which were at that time still airy and clean are now just as excessively built upon, just as dirty and over-crowded as the most ill-famed parts of the town formerly were. . . .

This is a striking example of how the bourgeoisie solves the housing question in practice. The breeding places of disease, the infamous holes and cellars in which the capitalist mode of production confines our workers night after night, are not abolished; they are merely *shifted else- where*! The same economic necessity which produced them in the first place, produces them in the next place also. As long as the capitalist mode of production continues to exist, it is folly to hope for an isolated solution of the housing question or of any other social question affect- ing the fate of the workers. The solution lies in the abolition of the cap- italist mode of production and the appropriation of all the means of life and labor by the working class itself.

Friedrich Engels, The Housing
Question, pp. 74-75, 77.

E. Conditions of Child Labor

Pillow lace-making is chiefly carried on in England in two agricultural districts; one, the Honiton lace district, extending from 20 to 30 miles along the south coast of Devonshire, and including a few places in North Devon; the other comprising a great part of the counties of Buckingham, Bedford, and Northampton, and also the adjoining portions of Oxford- shire and Huntingdonshire. The cottages of the agricultural labourers are the places where the work is usually carried on. Many manufacturers employ upwards of 3,000 of these lace-makers, who are chiefly children and young persons of the female sex exclusively. The state of things described as incidental to lace finishing is here repeated, save that instead of the "mistress' houses," we find what are called "lace-schools," kept by poor women in their cottages. From their fifth year and often earlier, until their twelfth or fifteenth year, the children work in these schools; during the first year the very young ones work from four to eight hours, and later on, from six in the morning till eight and ten o'clock at night. "The rooms are generally the ordinary living rooms of small cottages, the chimney stopped up to keep out draughts, the inmates kept warm by their own animal heat alone, and this frequently in winter. In other cases, these so-called school-rooms are like small store-rooms without fire-places. . . . The over-crowding in these dens and the consequent vitiation of the air are often extreme. Added to this is the injurious effect of drains, privies, decomposing substances, and other filth usual in the purlieus of the smaller cottages." With regard to space: "In one lace-school 18 girls and a mistress, 35 cubic feet to each person; in another, where the smell was unbearable, 18 persons and 24½ cubic feet per head. In this industry are to be found employed children of 2 and 2½ years."[1]

Where lace-making ends in the counties of Buckingham and Bedford, straw-plaiting begins, and extends over a large part of Hertfordshire and the westerly and northerly parts of Essex. In 1861, there were 40,043 persons employed in straw-plaiting and straw-hat making; of these 3,815 were males of all ages, the rest females, of whom 14,913, including about 7,000 children, were under 20 years of age. In the place of the lace-schools we find here the "straw-plait schools." The children commence their instruction in straw-plaiting generally in their 4th, often between their 3rd and 4th year. Education, of course, they get none. The children themselves call the elementary schools, "natural schools," to distinguish them from these blood-sucking institutions, in which they are kept at work simply to get through the task, generally 30 yards daily, prescribed by their half-starved mothers. These same mothers often make them work at home, after school is over, till 10, 11, and 12 o'clock at night. The straw cuts their mouths, with which they constantly moisten it, and their fingers. Dr. Ballard gives it as the general opinion of the whole body of medical officers in London, that 300 cubic feet is the minimum space proper for each person in a bedroom or workroom. But in the straw-plait schools space is more sparingly allotted than in the lace-schools, "12 2/3, 17, 18½ and below 22 cubic feet for each person." The smaller of these numbers, says one of the commissioners, Mr. White, represents less space than the half of what a child would occupy if packed in a box measuring 3 feet in each direction. Thus do the children enjoy life till the age of 12 or 14. The wretched half-starved parents think of nothing but getting as much as possible out of their children. The latter, as soon as they are grown up, do not care a farthing, and naturally so, for their parents, and leave them. "It is no wonder that ignorance and vice abound in a population so brought up. . . . Their morality is at the lowest ebb. . . . a great number of the women have illegitimate children, and that at such an immature age that even those most conversant with criminal statistics are astounded."[2] And the

---

1. "Ch. Empl. Comm., II. Rep., 1864," pp. xxix, xxx.
2. 1. c., pp. xl., xli.

native land of these model families is the pattern Christian country for
Europe; so says at least Count Montalembert, certainly a competent
authority on Christianity!

Wages in the above industries, miserable as they are (the maximum
wages of a child in the straw-plait schools rising in rare cases to 3 shillings),
are reduced far below their nominal amount by the prevalence of the
truck system everywhere, but especially in the lace districts.[3]

> Karl Marx, *Capital,* vol. 1, pp. 468-
> 469.

F. Pollution of the Air, Dwelling Place, Light, Clean Environment, Food, Activity, and Companionship of the Workers

This estrangement manifests itself in part in that it produces sophistica-
tion of needs and of their means on the one hand, and a bestial barbariza-
tion, a complete, unrefined, abstract simplicity of need, on the other; or
rather in that it merely resurrects itself in its opposite. Even the need for
fresh air ceases for the worker. Man returns to a cave dwelling, which is
now, however, contaminated with the pestilential breath of civilization,
and which he continues to occupy only *precariously,* it being for him
an alien habitation which can be withdrawn from him any day—a place
from which, if he does not pay, he can be thrown out any day. For this
mortuary he has to *pay.* A dwelling in the *light,* which Prometheus in
Aeschylus designated as one of the greatest boons, by means of which
he made the savage into a human being, ceases to exist for the worker.
Light, air, etc.—the simplest *animal* cleanliness—ceases to be a need for
man. *Filth,* this stagnation and putrefaction of man—the *sewage* of civilization (speaking quite literally)—comes to be the *element of life* for him.
Utter, *unnatural* neglect, putrefied nature, comes to be his *life-element.*
None of his senses exist any longer, and not only in his human fashion,
but in an *inhuman* fashion, and therefore not even in an animal fashion.
The crudest *methods* (and *instruments*) of human labor are coming back:
the *treadmill* of the Roman slaves, for instance, is the means of production, the means of existence, of many English workers. It is not only
that man has no human needs—even his *animal* needs cease to exist. The
Irishman no longer knows any need now but the need to *eat,* and indeed
only the need to eat *potatoes*—and *scabby potatoes* at that, the worst

---

3. "Child. Empl. Comm. I. Rep. 1863," p. 185.

kind of potatoes. But in each of their industrial towns England and France have already a *little* Ireland. The savage and the animal have at least the need to hunt, to roam, etc.—the need of companionship. Machine labor is simplified in order to make a worker out of the human being still in the making, the completely immature human being, the *child*—whilst the worker has become a neglected child. The machine accommodates itself to the *weakness* of the human being in order to make the *weak* human being into a machine.

> Karl Marx, *The Economic and Philosophic Manuscripts of 1844*, pp. 148-149.

X. The Transformation of Man's Relations to Nature Under Communism

Only through man's cooperation with the materials and forces of nature "as his own real body" does the way open for a real material basis for wealth and for the true wealth of human fulfillment. Such development, however, has occurred under a system that simultaneously obstructs the promised end. Marx states starkly the contrast between man's life under private property and man's life under communism, spelling out the difference in what happens to man, man's labor, and nature in each case. He foresaw in a grand and lyrical vision how labor can become liberating "as an activity regulating all the forces of nature." He foresaw how the processes of automation might be employed to reduce necessary labor time. He foresaw how, under socialist control and direction of society, and under man's rational cooperation with nature, all people might be released from backbreaking labor and the oppressions of the ruling class, and pass at last from the burdensome necessity of the past to true human freedom.

—Ed.

A. Under Capitalism the Development of the Productive Forces, While Encountering Contradictions, Creates the World Market, the Notion of History as Process, and the Recognition of Nature as Man's Body

Capital posits the *production of wealth* itself and hence the universal development of the productive forces, the constant overthrow of its prevailing presuppositions, as the presupposition of its reproduction. Value excludes no use value; i.e., includes no particular kind of consumption etc., of intercourse etc. as absolute condition; and likewise every degree of the development of the social forces of production, of intercourse, of knowledge etc. appears to it only as a barrier which it strives to overpower. Its own presupposition—value—is posited as product, not as a loftier presupposition hovering over production. The barrier to *capital* is that this entire development proceeds in a contradictory way, and that the working-out of the productive forces, of general wealth etc., knowledge etc., appears in such a way that the working individual *alienates* himself [*sich entäussert*] ; relates to the conditions brought out of him by his labour as those not of his *own* but of an *alien wealth* and of his own poverty. But this antithetical form is itself fleeting, and produces the real conditions of its own suspension. The result is: the tendentially and potentially general development of the forces of production—of wealth as such—as a basis; likewise, the universality of intercourse, hence the world market as a basis. The basis as the possibility of the universal development of the individual, and the real development of the individuals from this basis as a constant suspension of its *barrier,* which is recognized as a barrier, not taken for a *sacred limit.* Not an ideal or imagined universality of the individual, but the universality of his real and ideal relations. Hence also the grasping of his own history as a *process,* and the recognition of nature (equally present as practical power over nature) as his real body. The process of development itself posited and known as the presupposition of the same. For this, however, necessary above all that the full development of the forces of production has become the *condition of production*; and not that specific *conditions of production* are posited as a limit to the development of the productive forces.

Karl Marx, Grundrisse, pp. 541-542.

B. The Naturalism of Man and the Humanism of Nature

Re: p. XXXIX. The antithesis between *lack of property* and *property,* so long as it is not comprehended as the antithesis of *labor* and *capital,* still remains an indifferent antithesis, not grasped in its *active connection,* with its *internal* relation—an antithesis not yet grasped as a *contradiction.* It can find expression in this *first* form even without the advanced development of private property (as in ancient Rome, Turkey, etc.). It does not yet *appear* as having established by private property itself. But labor, the subjective essence of private property as exclusion of property, and capital objective labor as exclusion of labor, constitute *private property* as its developed state of contradiction—hence a dynamic relationship moving to its resolution.

Re: the same page. The transcendence of self-estrangement follows the same course as self-estrangement. *Private property* is first considered only in its objective aspect—but nevertheless with labor as its essence. Its form of existence is therefore *capital,* which is to be annulled "as such" (Proudhon). Or a *particular form* of labor—labor leveled down, parceled, and therefore unfree—is conceived as the source of private property's *perniciousness* and of its existence in estrangement from men. For instance, *Fourier,* who, like the physiocrats, also conceived *agricultural labor* to be at least the *exemplary* type, whilst *Saint-Simon* declares in contrast that *industrial labor* as such is the essence, and only aspires to the *exclusive* rule of the industrialists and the improvement of the workers' condition. Finally, *communism* is the *positive* expression of annulled private property—at first as *universal* private property. By embracing this relation as a *whole,* communism is:

(1) In its first form only a *generalization* and *consummation* of this relationship. As such it appears in a twofold form: on the one hand, the dominion of *material* property bulks so large that it wants to destroy *everything* which is not capable of being possessed by all as *private property.* It wants to do away *by force* with talent, etc. For it the sole purpose of life and existence is direct, physical *possession.* The task of the *laborer* is not done away with, but extended to all men. The relationship of private property persists as the relationship of the community to the world of things.

Finally, this movement of opposing universal private property to private property finds expression in the animal form of opposing to *marriage* (certainly a *form of exclusive private property*) the *community of women,* in which a woman becomes a piece of *communal* and *common* property. It may be said that this idea of the *community of women* gives away the *secret* of this as yet completely crude and thoughtless communism. Just as woman passes from marriage to general prostitution,[1] so the entire

---

1. Prostitution is only a *specific* expression of the *general* prostitution of the *laborer,* and since it is a relationship in which falls not the prostitute alone, but also the one who prostitutes—and the latter's abomination is still greater—the capitalist, etc., also comes under this head.

world of wealth (that is, of man's objective substance) passes from the relationship of exclusive marriage with the owner of private property to a state of universal prostitution with the community. In negating the *personality* of man in every sphere, this type of communism is really nothing but the logical expression of private property, which is its negation. General *envy* constituting itself as a power is the disguise in which *greed* reestablishes itself and satisfies itself, only in another way. The thought of every piece of private property—inherent in each piece as such— is *at least* turned against all *wealthier* private property in the form of envy and the urge to reduce things to a common level, so that this envy and urge even constitute the essence of competition. The crude communism is only the culmination of this envy and of this leveling-down proceeding from the *preconceived* minimum. It has a *definite, limited* standard. How little this annulment of private property is really an appropriation is in fact proved by the abstract negation of the entire world of culture and civilization, the regression to the *unnatural* simplicity of the *poor and undemanding* man who has not only failed to go beyond private property, but has not yet even reached it.

The community is only a community of *labor,* and of equality of *wages* paid out by communal capital—the *community* as the universal capitalist. Both sides of the relationship are raised to an *imagined* universality—*labor* as a state in which every person is placed, and *capital* as the acknowledged universality and power of the community.

In the approach to *woman* as the spoil and handmaid of communal lust is expressed the infinite degradation in which man exists for himself, for the secret of this approach has its *unambiguous,* decisive, *plain* and undisguised expression in the relation of *man* to *woman* and in the manner in which the *direct* and *natural* species relationship is conceived. This direct, natural, and necessary relation of person to person is the *relation of man to woman.* In this *natural* species relationship man's relation to nature is immediately his relation to man, just as his relation to man is immediately his relation to nature—his own *natural* destination. In this relationship, therefore, is *sensuously manifested,* reduced to an observable *fact,* the extent to which the human essence has become nature to man, or to which nature to him has become the human essence of man. From this relationship one can therefore judge man's whole level of development. From the character of this relationship follows how much *man* as a *species being,* as *man,* has come to be himself and to comprehend himself;

the relation of man to woman is *the most natural* relation of human being to human being. It therefore reveals the extent to which man's *natural* behavior has become *human,* or the extent to which the *human* essence in him has become a *natural* essence—the extent to which his *human nature* has come to be *nature to him.* In this relationship is revealed, too, the extent to which man's *need* has become a *human* need; the extent to which, therefore, the *other* person as a person has become for him a need—the extent to which he in his individual existence is at the same time a social being.

The first positive annulment of private property—*crude* communism— is thus merely one *form* in which the vileness of private property, which wants to set itself up as the *positive community, comes to the surface.*

(2) Communism (*a*) still political in nature—democratic or despotic; (*b*) with the abolition of the state, yet still incomplete, and being still affected by private property (i.e., by the estrangement of man). In both forms communism already is aware of being reintegration or return of man to himself, the transcendence of human self-estrangement; but since it has not yet grasped the positive essence of private property, and just as little the *human* nature of need, it remains captive to it and infected by it. It has, indeed, grasped its concept, but not its essence.

(3) *Communism* as the *positive* transcendence of *private property,* as *human self-estrangement,* and therefore as the real *appropriation of the human* essence by and for man; communism therefore as the complete return of man to himself as a *social* (i.e., human) being—a return become conscious, and accomplished within the entire wealth of previous develop- ment. This communism, as fully developed naturalism, equals humanism, and as fully developed humanism equals naturalism; it is the *genuine* resolution of the conflict between man and nature and between man and man—the true resolution of the strife between existence and essence, between objectification and self-confirmation, between freedom and necessity, between the individual and the species. Communism is the riddle of history solved, and it knows itself to be this solution.

The entire movement of history is, therefore, both its *actual* act of genesis (the birth act of its empirical existence) and also for its thinking consciousness the *comprehended* and *known* process of its *becoming.* That other, still immature communism, meanwhile, seeks an *historical* proof for itself—a proof in the realm of what already exists—among dis- connected historical phenomena opposed to private property, tearing single phases from the historical process and focusing attention on them as proofs of its historical pedigree (a hobbyhorse ridden hard especially by Cabet, Villegardelle, etc.). By so doing it simply makes clear that by far the greater part of this process contradicts its own claim, and that, if it has ever existed, precisely its being in the *past* refutes its pretension to being *essential being.*

It is easy to see that the entire revolutionary movement necessarily finds both its empirical and its theoretical basis in the movement of *private property*—more precisely, in that of the economy.

This *material,* immediately perceptible private property is the material perceptible expression of *estranged human* life. Its movement—production and consumption—is the *perceptible* revelation of the movement of all production until now, i.e., the realization or the reality of man. Religion, family, state, law, morality, science, art, etc., are only *particular* modes of production, and fall under its general law. The positive transcendence of *private property,* as the appropriation of *human* life, is therefore the positive transcendence of all estrangement—that is to say, the return of man from religion, family, state, etc., to his *human,* i.e., *social* existence. Religious estrangement as such occurs only in the realm of *consciousness,* of man's inner life, but economic estrangement is that of *real life;* its transcendence therefore embraces both aspects. It is evident that the *initial* stage of the movement amongst the various peoples depends on whether the true and *authentic* life of the people manifests itself more in consciousness or in the external world—is more ideal or real. Communism begins from the outset (*Owen*) with atheism; but atheism is at first far from being *communism*; indeed it is still mostly an abstraction.

The philanthropy of atheism is therefore at first only *philosophical,* abstract, philanthropy, and that of communism is at once *real* and directly bent on *action.*

We have seen how on the assumption of positively annulled private property man produces man—himself and the other man; how the object, being the direct embodiment of his individuality, is simultaneously his own existence for the other man, the existence of the other man, and that existence for him. Likewise, however, both the material of labor and man as the subject, are the point of departure as well as the result of the movement (and precisely in this fact, that they must constitute the *point of departure,* lies the historical *necessity* of private property). Thus the *social* character is the general character of the whole movement; *just as* society itself produces *man as man,* so is society *produced* by him. Activity and mind, both in their content and in their *mode of existence,* are *social: social* activity and *social* mind. The *human* essence of nature first exists only for *social* man; for only here does nature exist for him as a *bond* with *man*—as his existence for the other and the other's existence for him—as the life-element of human reality. Only here does nature exist as the *foundation* of his own *human* existence. Only here has what is to him his *natural* existence become his *human* existence, and nature become man for him. Thus *society* is the unity of being of man with nature—the true resurrection of nature—the naturalism of man and the humanism of nature both brought to fulfillment.

Social activity and social mind exist by no means *only* in the form of some *directly* communal activity and directly *communal* mind, although *communal* activity and *communal* mind—i.e., activity and mind which are manifested and directly revealed in *real association* with other men—will occur wherever such a *direct* expression of sociability stems from the true character of the activity's content and is adequate to its nature.

But also when I am active *scientifically,* etc.—when I am engaged in activity which I can seldom perform in direct community with others—then I am *social,* because I am active as a *man.* Not only is the material of my activity given to me as a social product (as is even the language in which the thinker is active): my *own* existence *is* social activity, and therefore that which I make of myself, I make of myself for society and with the consciousness of myself as a social being.

My *general* consciousness is only the *theoretical* shape of that which the *living* shape is the *real* community, the social fabric, although at the present day *general* consciousness is an abstraction from real life and as such confronts it with hostility. The *activity* of my general consciousness, as an activity, is therefore also my *theoretical* existence as a social being.

Above all we must avoid postulating "Society" again as an abstraction *vis-à-vis* the individual. The individual *is the social being.* His life, even if it may not appear in the direct form of a *communal* life in association with others—is therefore an expression and confirmation of *social life.* Man's individual and species life are not *different,* however much—and this is inevitable—the mode of existence of the individual is a more *particular,* or more *general* mode of the life of the species, or the life of the species is a more *particular* or more *general* individual life.

In his *consciousness of species* man confirms his real *social life* and

simply repeats his real existence in thought, just as conversely the being
of the species confirms itself in species-consciousness and exists for *itself*
in its generality as a thinking being.

Man, much as he may therefore be a *particular* individual (and it is
precisely his particularity which makes him an individual, and a real *in-dividual* social being), is just as much the *totality*—the ideal totality—the
subjective existence of thought and experienced society for itself; just
as he exists also in the real world as the awareness and the real mind of
social existence, and as a totality of human manifestation of life.

Thinking and being are thus no doubt *distinct*, but at the same time
they are in *unity* with each other.

Death seems to be a harsh victory of the species over the *definite*
individual and to contradict their unity. But the particular individual is
only a *particular species being,* and as such mortal.

(4) Just as *private property* is only the perceptible expression of the
fact that man becomes *objective* for himself and at the same time becomes
to himself a strange and inhuman object; just as it expresses the fact that
the assertion of his life is the alienation of his life, that his realization is
his loss of reality, is an *alien* reality: so, the positive transcendence of
private property—i.e., the *perceptible* appropriation for and by man of
the human essence and of human life, of objective man, of human *achieve-ments*—should not to be conceived merely in the sense of *immediate,*
one-sided *gratification*—merely in the sense of *possessing,* of *having.* Man
appropriates his total essence in a total manner, that is to say, as a whole
man. Each of his *human* relations to the world—seeing, hearing, smelling,
tasting, feeling, thinking, observing, experiencing, wanting, acting, loving—
in short, all the organs of his individual being, like those organs which
are directly social in their form, are in their *objective* orientation or in
their *orientation to the object,* the appropriation of that object. The
appropriation of *human* reality;[2] its orientation to the object is the *man-ifestation of the human reality,* it is human *activity* and human *suffering,*
for suffering, humanly considered, is a self-indulgence of man.

Private property has made us so stupid and one-sided that an object is
only *ours* when we have it—when it exists for us as capital, or when it is
directly possessed, eaten, drunk, worn, inhabited, etc.—in short, when it

---

2. For this reason it is just as highly varied as the *determinations* of human
essence and *activities.*

is *used* by us. Although private property itself again conceives all these direct realizations of possession only as *means of life,* and the life which they serve as means is the *life of private property*—labor and conversion into capital.

All these physical and mental senses have therefore—the sheer estrangement of *all* these senses—the sense of *having.* The human being had to be reduced to this absolute poverty in order that he might yield his inner wealth to the outer world. (On the category of *"having,"* see Hess in the *Twenty-One Sheets.*)

The transcendence of private property is therefore the complete *emancipation* of all human senses and qualities, but it is this emancipation precisely because these senses and attributes have become, subjectively and objectively, *human.* The eye has become a *human* eye, just as its *object* has become a social, *human* object—an object made by man for man. The *senses* have therefore become directly in their practice *theoreticians.* They relate themselves to the *thing* for the sake of the thing, but the thing itself is an *objective human* relation to itself and to man,[3] and vice versa. Need or enjoyment have consequently lost their *egotistical* nature, and nature has lost its mere *utility* by use becoming *human* use.

In the same way, the senses and minds of other men have become my *own* appropriation. Besides these direct organs, therefore, *social* organs develop in the *form* of society; thus, for instance, activity in direct association with others, etc., has become an organ for *expressing* my own *life,* and a mode of appropriating *human* life.

It is obvious that the *human* eye enjoys things in a way different from the crude, non-human eye; the human *ear* different from the crude ear, et

To recapitulate: man is not lost in his object only when the object becomes for him a *human* object or objective man. This is possible only when the object becomes for him a *social* object, he himself for himself a social being, just as society becomes a being for him in this object.

On the one hand, therefore, it is only when the objective world become everywhere for man in society the world of man's essential powers—huma reality, and for that reason the reality of his *own* essential powers—that all *objects* become for him the *objectification of himself,* become objects which confirm and realize his individuality, become *his* objects: that is, *man himself* becomes the object. The manner in which they become *his*

---

3. In practice I can relate myself to a thing humanly only if the thing relates itself humanly to the human being.

depends on the *nature of the objects* and on the nature of the *essential power* corresponding *to it;* for it is precisely the *determinate nature* of this relationship which shapes the particular, *real* mode of affirmation. To the *eye* an object comes to be other than it is to the *ear,* and the object of the eye is another object than the object of the *ear.* The specific character of each essential power is precisely its *specific essense,* and therefore also the specific mode of its objectification, of its *objectively actual* living *being.* Thus man is affirmed in the objective world not only in the act of thinking, but with *all* his senses.

On the other hand, let us look at this in its subjective aspect. Just as music alone awakens in man the sense of music, and just as the most beautiful music has *no* sense for the unmusical ear—is no object for it, because my object can only be the confirmation of one of my essential powers. It can therefore only be so for me as my essential power exists for itself as a subjective capacity, because the meaning of an object for me goes only so far as *my* senses go (has only a meaning for a sense corresponding to that object)—for this reason the *senses* of the social man are *other* senses than those of the non-social man. Only through the objectively unfolded richness of man's essential being is the richness of subjective *human* sensibility (a musical ear, an eye for beauty of form—in short, *senses* capable of human gratification, senses affirming themselves as essential powers of *man*) either cultivated or brought into being. For not only the five senses but also the so-called mental senses—the practical senses (will, love, etc.)—in a word, *human* sense—the human nature of the senses—comes to be by virtue of its object, by virtue of *humanized* nature. The *forming* of the five senses is a labor of the entire history of the world down to the present.

The *sense* caught up in crude practical need has only a restricted sense. For the starving man, it is not the human form of food that exists, but only its abstract being as food. It could just as well be there in its crudest form, and it would be impossible to say wherein this feeding activity differs from that of *animals.* The care-burdened man in need has no sense for the finest play; the dealer in minerals sees only the commercial value but not the beauty and the unique nature of the mineral: he has no mineralogical sense. Thus, the objectification of the human essence, both in its theoretical and practical aspects, is required to make man's *sense human,* as well as to create the *human sense* corresponding to the entire wealth of human and natural substance.

Just as through the movement of *private property,* of its wealth as

well as its poverty—or of its material and spiritual wealth and poverty—
the budding society finds at hand all the material for this *development,*
so *established* society produces man in this entire richness of his being—
produces the *rich* man *profoundly endowed with all the senses*—as its
enduring reality.

We see how subjectivism and objectivism, spiritualism and materialism,
activity and suffering, only lose their antithetical character, and thus their
existence as such antitheses in social centers; we see how the resolution
of the *theoretical* antitheses is *only* possible *in a practical* way, by virtue
of the practical energy of man. Their resolution is therefore by no means
merely a problem of understanding, but a *real* problem of life, which
philosophy could not solve precisely because it conceived this problem
as *merely* a theoretical one.

We see how the history of *industry* and the established *objective* existence of industry are the *open book* of *man's essential powers,* the exposure to the senses of human *psychology.* Hitherto this was not conceived
in its inseparable connection with man's *essential being,* but only in an
external relation of utility, because, moving in the realm of estrangement,
people could only think of man's general mode of being—religion or history
in its abstract-general character as politics, art, literature, etc.—as the reality of man's essential powers and *man's species activity.* We have before
us the *objectified essential powers* of man in the form of *sensuous, alien,
useful objects,* in the form of estrangement, displayed in *ordinary material
industry* (which can be conceived as well as a part of that general movement, just as that movement can be conceived as a *particular* part of
industry, since all human activity hitherto has been labor—that is, industry—
activity estranged from itself).

A *psychology* for which this, the part of history most contemporary
and accessible to sense, remains a closed book, cannot become a genuine,
comprehensive and *real* science. What indeed are we to think of a science
which *airily* abstracts from this large part of human labor and which fails
to feel its own incompleteness, while such a wealth of human endeavor,
unfolded before it, means nothing more to it than, perhaps, what can be
expressed in one word—*"need," "vulgar need"*?

The *natural sciences* have developed an enormous activity and have
accumulated an ever-growing mass of material. Philosophy, however, has
remained just as alien to them as they remain to philosophy. Their momentary unity was only a *chimerical illusion.* The will was there, but the
means were lacking. Even historiography pays regard to natural science

only occasionally, as a factor of enlightenment, utility, and of some special great discoveries. But natural science has invaded and transformed human life all the more *practically* through the medium of industry; and has prepared human emancipation, although its immediate effect had to be the furthering of the dehumanization of man. *Industry* is the *actual,* historical relationship of nature, and therefore of natural science, to man. If, therefore, industry is conceived as the *exoteric* revelation of man's *essential powers,* we also gain an understanding of the *human* essence of nature or the *natural* essence of man. In consequence, natural science will lose its abstractly material—or rather, its idealistic—tendency, and will become the basis of *human* science, as it has already become the basis of actual human life, albeit in an estranged form. *One* basis for life and another basis for *science* is *a priori* a lie. The nature which develops in human history—the genesis of human society—is man's *real* nature; hence nature as it develops through industry, even though in an *estranged* form, is true *anthropological* nature.

Sense-perception (see Feuerbach) must be the basis of all science. Only when it proceeds from sense-perception in the two-fold form both of *sensuous* consciousness and of *sensuous* need—that is, only when science proceeds from nature—is it *true* science. All history is the preparation for *"man"* to become the object of *sensuous* consciousness, and for the needs of "man as man" to become [natural, sensuous] needs. History itself is a *real* part of *natural history*—of nature developing into man. Natural science will in time incorporate into itself the science of man, just as the science of man will incorporate into itself natural science: there will be *one* science.

Man is the immediate object of natural science; for immediate, *sensuous nature* for man is, immediately, human sensuousness (the expressions are identical)—presented immediately in the form of the *other* man sensuously present for him. Indeed, his own sensuousness first exists as human sensuousness for himself through the *other* man. But *nature* is the immediate object of the *science of man;* the first object of man—man—is nature, sensuousness; and the particular sensuous human essential powers can only find their self-understanding in the science of the natural world in general, since they can find their objective realization in *natural* objects only. The element of thought itself—the element of thought's living expression—*language*—is of a sensuous nature. The *social* reality of nature, and *human* natural science, or the *natural science about man,* are identical terms.

It will be seen how in place of the *wealth* and *poverty* of political economy comes the *rich human being* and the rich *human* need. The *rich* human being is simultaneously the human being *in need of* a totality of human manifestations of life—the man in whom his own realization exists as an inner necessity, as *need.* Not only *wealth,* but likewise the *poverty* of man—under the assumption of socialism—receives in equal measure a *human* and therefore social significance. Poverty is the passive bond which causes the human being to experience the need of the greatest wealth—the *other* human being. The dominion of the objective being in me, the sensuous outburst of my life activity, is *passion,* which thus becomes here the *activity* of my being.

(5) A *being* only considers himself independent when he stands on his own feet; and he only stands on his own feet when he owes his *existence* to himself. A man who lives by the grace of another regards himself as a dependent being. But I live completely by the grace of another if I owe him not only the maintenance of my life, but if he has, moreover, *created* my *life*—if he is the *source* of my life. When it is not of my own creation, my life has necessarily a source of this kind outside of it. The *Creation* is therefore an idea very difficult to dislodge from popular consciousness. The fact that nature and man exist in their own account is *incomprehensible* to it, because it contradicts everything *tangible* in practical life.

The creation of the *earth* has received a mighty blow from geogeny—i.e., from the science which presents the formation of the earth, the further development of the earth, as a process, as a self-generation. *Generatio aequivoca* is the only practical refutation of the theory of creation.

Now it is certainly easy to say to the single individual what Aristotle has already said: You have been begotten by your father and your mother; therefore in you the mating of two human beings—a species-act of human beings—has produced the human being. You see, therefore, that even physically, man owes his existence to man. Therefore you must not only keep sight of the *one* aspect—the *infinite* progression which leads you further to enquire: "Who begot my father? Who his grandfather?," etc. You must also hold on to the *circular movement* sensuously perceptible in that progression, by which *man* repeats himself in procreation, *man* thus always remaining the subject. You will reply, however: I grant you this circular movement; now grant me the progression which drives me ever further until I ask: Who begot the first man, and nature as a whole? I can only answer you: Your question is itself a product of abstraction. Ask yourself how you arrived at that question. Ask yourself whether your question is not posed from a standpoint to which I cannot reply, because it is wrongly put. Ask yourself whether that progression as such exists for a reasonable mind. When you ask about the creation of nature and man, you are abstracting, in so doing, from man and nature. You postulate them as *non-existent*, and yet you want me to prove them to you as *existing*. Now I say to you: Give up your abstraction and you will also give up your question. Or if you want to hold on to your abstraction, then be consistent, and if you think of man and nature as *non-existent*, then think of yourself as non-existent, for you too are surely nature and man. Don't think, don't ask me, for as soon as you think and ask, your *abstraction* from the existence of nature and man has no meaning. Or are you such an egotist that you conceive everything as nothing, and yet want yourself to exist?

You can reply: I do not want to conceive the nothingness of nature, etc. I ask you about its *genesis*, just as I ask the anatomist about the formation of bones, etc.

But since for the socialist man the *entire so-called history of the world* is nothing but the creation of man through human labor, nothing but the emergence of nature for man, so he has the visible, irrefutable proof of his *birth* through himself, of the *process of his creation*. Since the *real existence* of man and nature—since man has become for man as the being of nature, and nature for man as the being of man has become practical, sensuous, perceptible—the question about an *alien* being, about a being above nature and man—a question which implies the admission of the unreality of nature and of man—has become impossible in practice. *Atheism*, as the denial of this unreality, has no longer any meaning, for atheism is a *negation of God*, and postulates the *existence of man* through this negation; but socialism as socialism no longer stands in any need of such a mediation. It proceeds from the *practically and theoretically sensuous consciousness* of man and of nature as the *essence*. Socialism is man's *positive self-consciousness*, no longer mediated through the annulment of religion, just as *real life* is man's positive reality, no longer mediated through the annulment of private property, through *communism*. Communism is the position as the negation of the negation, and is hence the *actual* phase necessary for the next stage of historical development in the process of human emancipation and rehabilitation. *Communism* is the necessary pattern and the dynamic principle of the immediate future, but communism as such is not the goal of human development—which goal is the structure of human society.

> Karl Marx, *The Economic and Philosophic Manuscripts of 1844*, pp. 132-146.

C. Man's Labor Is Liberated When It Becomes Social and Universally Regulative of Nature

In the sweat of thy brow shalt thou labour! was Jehovah's curse on Adam. And this is labour for Smith, a curse. 'Tranquility' appears as the inadequate state, as identical with 'freedom' and 'happiness.' It seems quite far from Smith's mind that the individual, 'in his normal state of health, strength, activity, skill, facility,' also needs a normal portion of work, and of the suspension of tranquility. Certainly, labour obtains its measure from the outside, through the aim to be attained and the obstacles to be overcome in attaining it. But Smith has no inkling whatever that this overcoming of obstacles is in itself a liberating activity—and that, further, the external aims become stripped of the semblance of merely external natural urgencies, and become posited as aims which the individual himself posits—hence as self realization, objectification of the subject, hence real freedom, whose action is, precisely, labour. He is right, of course, that, in its historic forms as slave-labour, serf-labour, and wage-labour, labour always appears as repulsive, always as *external forced labour*; and not-labour, by contrast, as 'freedom, and happiness.' This holds doubly: for this contradictory labour; and, relatedly, for labour which has not yet created the subjective and objective conditions for itself (or also, in contrast to the pastoral etc. state, which it has lost), in which labour becomes attractive work, the individual's self-realization, which in no way means that it becomes mere fun, mere amusement, as Fourier, with *grisette*-like naïvete, conceives it. Really free working, e.g. composing, is at the same time precisely the most damned seriousness, the most intense exertion. The work of material production can achieve this character only (1) when its social character is posited, (2) when it is of a scientific and at the same time general character, not merely human exertion as a specifically harnessed natural force, but exertion as subject, which appears in the production process not in a merely natural, spontaneous form, but as an activity regulating all the forces of nature.

> Karl Marx, *Grundrisse*, pp. 611-612.

*D. Man's Knowledge and Mastery of Natural Processes to Develop Auto-
mation, as the New Basis of Social Wealth, Contradicts Capitalism's
Outmoded Exploitation of Labor Time*

The exchange of living labour for objectified labour—i.e. the positing
of social labour in the form of the contradiction of capital and wage labour—
is the ultimate development of the *value-relation* and of production resting on value. Its presupposition is—and remains—the mass of direct labour
time, the quantity of labour employed, as the determinant factor in the
production of wealth. But to the degree that large industry develops, the
creation of real wealth comes to depend less on labour time and on the
amount of labour employed than on the power of the agencies set in motion during labour time, whose 'powerful effectiveness' is itself in turn
out of all proportion to the direct labour time spent on their production,
but depends rather on the general state of science and on the progress of
technology, or the application of this science to production. (The development of this science, especially natural science, and all others with the
latter, is itself in turn related to the development of material production.)
Agriculture, e.g., becomes merely the application of the science of material
metabolism, its regulation for the greatest advantage of the entire body
of society. Real wealth manifests itself, rather—and large industry reveals
this—in the monstrous disproportion between the labour time applied,
and its product, as well as in the qualitative imbalance between labour,
reduced to a pure abstraction, and the power of the production process
it superintends. Labour no longer appears so much to be included within
the production process; rather, the human being comes to relate more as
watchman and regulator to the production process itself. (What holds for
machinery holds likewise for the combination of human activities and
the development of human intercourse.) No longer does the worker insert a modified natural thing [*Naturgegenstand*] as middle link between
the object [*Objekt*] and himself; rather, he inserts the process of nature,
transformed into an industrial process, as a means between himself and
and inorganic nature, mastering it. He steps to the side of the production
process instead of being its chief actor. In this transformation, it is neither
the direct human labour he himself performs, nor the time during which
he works, but rather the appropriation of his own general productive
power, his understanding of nature and his mastery over it by virtue of
his presence as a social body—it is, in a word, the development of the
social individual which appears as the great foundation-stone of produc-

tion and of wealth. The *theft of alien labour time, on which the present wealth is based,* appears a miserable foundation in face of this new one, created by large-scale industry itself. As soon as labour in the direct form has ceased to be the great well-spring of wealth, labour time ceases and must cease to be its measure, and hence exchange value [must cease to be the measure] of use value. The *surplus labour of the mass* has ceased to be the condition for the development of general wealth, just as the *non-labour of the few,* for the development of the general powers of the human head. With that, production based on exchange value breaks down, and the direct, material production process is stripped of the form of penury and antithesis. The free development of individualities, and hence not the reduction of necessary labour time so as to posit surplus labour, but rather the general reduction of the necessary labour of society to a minimum, which then corresponds to the artistic, scientific etc. development of the individuals in the time set free, and with the means created, for all of them. Capital itself is the moving contradiction, [in] that it presses to reduce labour time to a minimum, while it posits labour time, on the other side, as sole measure and source of wealth. Hence it diminishes labour time in the necessary form so as to increase it in the superfluous form; hence posits the superfluous in growing measure as a condition—question of life or death—for the necessary. On the one side, then, it calls to life all the powers of science and of nature, as of social combination and of social intercourse, in order to make the creation of wealth independent (relatively) of the labour time employed on it. On the other side, it wants to use labour time as the measuring rod for the giant social forces thereby created, and to confine them within the limits required to maintain the already created value as value. Forces of production and social relations—two different sides of the development of the social individual—appear to capital as mere means, and are merely means for it to produce on its limited foundation. In fact, however, they are the material conditions to blow this foundation sky-high. 'Truly wealthy a nation, when the working day is 6 rather than 12 hours. *Wealth* is not command over surplus labour time' (real wealth), 'but rather, *disposable time* outside that needed in direct production, for *every individual* and the whole society.' (*The Source and Remedy* etc. 1821, p. 6.)

    Nature builds no machines, no locomotives, railways, electric telegraphs, self-acting mules etc. These are products of human industry; natural material transformed into organs of the human will over nature, or of human participation in nature. They are *organs of the human brain, created by the human hand;* the power of knowledge, objectified. The development of fixed capital indicates to what degree general social knowledge has become a *direct force of production,* and to what degree, hence the conditions of the process of social life itself have come under the control of the general intellect and been transformed in accordance with it. To what degree the powers of social production have been produced, not only in the form of knowledge, but also as immediate organs of social practice, of the real life process.

Karl Marx, Grundrisse, pp. 704-706.

E. Labor's Productivity and the Conditions of Production, Not Surplus Labor Time, Create Wealth; But Only When Socialized Men Rationally Regulate Their Exchanges with Nature Is a Basis Laid in Necessity for True Human Freedom

The actual wealth of society, and the possibility of constantly expanding its reproduction process, therefore, do not depend upon the duration of surplus-labour, but upon its productivity and the more or less copious conditions of production under which it is performed. In fact, the realm of freedom actually begins only where labour which is determined by necessity and mundane considerations ceases; thus in the very nature of things it lies beyond the sphere of actual material production. Just as the savage must wrestle with Nature to satisfy his wants, to maintain and reproduce life, so must civilised man, and he must do so in all social formations and under all possible modes of production. With his development this realm of physical necessity expands as a result of his wants; but, at the same time, the forces of production which satisfy these wants also increase. Freedom in this field can only consist in socialised man, the associated producers, rationally regulating their interchange with Nature, bringing it under their common control, instead of being ruled by it as by the blind forces of Nature; and achieving this with the least expenditure of energy and under conditions most favourable to, and worthy of, their human nature. But it nonetheless still remains a realm of necessity. Beyond it begins that development of human energy which is an end in itself, the true realm of freedom, which, however, can blossom forth only with this realm of necessity as its basis. The shortening of the working-day is its basic prerequisite.

Karl Marx, Capital, vol. 3, p. 820.

Bibliography

Abramov, L. S., et al., eds. *The Interaction of Nature and Society. Philosophical, Geographical, and Ecological Aspects of the Problem.* Moscow, 1973. (In Russian.)

Abrams, Charles. *The City Is the Frontier.* New York: Harper and Row, 1965.

——. *Man's Struggle in an Urbanizing World.* Cambridge, Mass.: M.I.T. Press, 1966.

Ackoff, Russell L. *Redesigning the Future: A Systems Approach to Societal Problems.* New York: John Wiley and Sons, 1974.

Adabashev, I. I. *Tragedy or Harmony? Nature-Machine-Man.* Moscow: Mysl, 1973. (In Russian.)

Adelstein, Michael E., and Pival, Joan G. *Ecocide and Population.* New York: St. Martin's Press, 1972.

Albertson, Peter, and Barnett, Margery. *Managing the Planet.* Englewood Cliffs, N.J.: Prentice-Hall, 1972.

Allee, W. C.; Emerson, Alfred; Park, Orlando; Park, Thomas; and Schmidt, Karl P., eds. *Principles of Animal Ecology.* Philadelphia: W. B. Saunders, 1949.
An early and definitive classic in the field; both a survey and an evaluation of the literature.

Allen, Jonathan, ed. *Scientists, Students, and Society.* Cambridge, Mass.: M.I.T. Press, 1970.
Various scientists on the scientist's responsibility toward society and government on issues such as war, disarmament, and weapons research.

Allsopp, Bruce. *Ecological Morality.* London: Frederick Muller, 1972.

American Association for the Advancement of Science. *Air Conservation. The Classic Study.* Washington, D.C., 1973. Based on the 1973 AAAS symposium.

American Chemical Society. *Chemistry and the Environment: The Solid Earth, the Oceans, the Atmosphere.* Washington, D.C., 1967.

——. *Cleaning Our Environment: The Chemical Basis for Action.* Washington, D.C., 1969.
On air, water, solid wastes, and pesticides. Copious literature cited.

Anderson, Paul K., ed. *Omega: Murder of the Ecosystem and Suicide of Man.* Dubuque, Iowa: William C. Brown, 1971.
"The ecological childhood of man is over, and it has ended without the gift

of ecological wisdom." The reason, according to the editor, is that human biology, culture, and technology are "growth adapted," and relative to population are in conflict with man's asymptotic relation to earth's carrying capacity.

Anderson, Walt, ed. *Politics and Environment: A Reader in Ecological Crisis.* Pacific Palisades: Goodyear, 1970.

Andrews, William A.; Moore, Donna K.; and LeRoy, Alex C. *A Guide to the Study of Environmental Pollution.* Englewood Cliffs, N.J.: Prentice-Hall, 1972.

Ariyoshi, S. *Fukugo-osen [Complex pollution].* 2 vols. Tokyo: Shinchosa, 1975. (In Japanese.)

Arvill, Robert. *Man and Environment: Crisis and the Strategy of Choice.* Harmondsworth, England: Penguin Books, 1973.

Ashford, Nicholas A. *Crisis in the Workplace: Occupational Disease, and Injury.* A Report to the Ford Foundation. Cambridge, Mass.: M.I.T. Press, 1975. Demonstration of the failure of management in U.S. industry to deal with problems of safety and health. Includes an expose' of the economic forces opposed to the standards of the Occupational Safety and Health Administration and a study of the safety and health programs in Western European countries, superior to those in the United States.

Asimov, Isaac. *Earth: Our Crowded Spaceship.* New York: John Day, 1974.

Baade, Fritz. *The Race to the Year 2000,* trans. Ernst Pawel. Garden City, N.Y.: Doubleday, 1962.

Bachelard, Gaston. *The Poetics of Space.* Trans. Maria Jolas. Boston: Beacon Press, 1969. Poetic and penetrating exposition of the meaning and value of the space that we inhabit and that inhabits us—"the space we love." See ch. 1, "The House. From Cellar to Garret. The Significance of the Hut."

Baier, Kurt, and Rescher, Nicholas, eds. *Values and the Future: The Impact of Technological Change on American Values.* New York: Free Press, 1969.

Baran, Paul A. *The Political Economy of Growth.* New York: Monthly Review Press, 1957. Devastating diagnosis of the stagnation of monopoly capitalism in advanced and backward countries, and its self-defeating investment of surplus (vital to growth) through government spending, militarism, unproductive labor, and imperialism.

——, and Sweezy, Paul M. *Monopoly Capital.* New York: Monthly Review Press, 1966. The giant corporation's problem of utilizing rising surplus by private consumption and investment, sales, civilian government, militarism, and imperialism.

Barbour, Ian G. *Earth Might Be Fair: Reflections on Ethics, Religion and Ecology.* Englewood Cliffs, N.J.: Prentice-Hall, 1972. A philosopher competent in both science and religion reflects on ecological issues.

Barr, John. *Derelict Britain.* Harmondsworth, England: Penguin, 1969.

Bates, Marston. *Man in Nature.* 2d ed. Englewood Cliffs, N.J.: Prentice-Hall, 1964.

Wide-ranging zoologist and ecologist.

Bateson, Gregory. *Steps to an Ecology of Mind*. San Francisco: Chandler Publishing Co., 1972.

An assembly of the results of thirty-five years of exploration into the conditions, interaction, selection, and survival of ideas and systems of ideas (minds).

Baxter, William F. *People or Penguins? The Case for Optimal Pollution*. New York: Columbia University Press, 1974.

Behrman, Daniel. *In Partnership with Nature*. Paris: UNESCO, 1972.

Bernal, J. D. *Science in History*. 4 vols. London: C. A. Watts, 1954.

Like most of Bernal's work, a pioneer achievement of durable value.

——. *The Social Function of Science*. London: Routledge and Kegan Paul, 1939; Cambridge, Mass.: M.I.T. Press, 1967.

The first full study of its kind—and from a Marxist perspective.

Berrill, N. J. *Man's Emerging Mind*. New York: Dodd, Mead, 1955.

Imaginative and scientific reconstruction of the primate origins of man.

Biolat, G. *Marxisme et environnement*. Paris: Ed. sociales, 1973.

Black, John. *Dominion of Man: The Search for Ecological Responsibility*. Chicago: Aldine Publishing Co., 1970.

Blackstone, William T., ed., *Philosophy and Environmental Crisis*. Athens: University of Georgia Press, 1974.

Philosophers respond to issues raised by Eugene Odum's essay calling for an "environmental ethic."

Blake, Peter. *God's Own Junkyard: The Planned Deterioration of America's Landscape*. New York: Holt, Rinehart and Winston, 1964.

Anticipates the muckraking of the 1970s.

Bonner, James. *The Next Ninety Years*. Pasadena: California Institute of Technology, 1967.

Bookchin, Murray, and Ecology Action East. *Ecology and Revolutionary Thought*. New York: Times Change, 1971.

An anarchist viewpoint.

Borgstrom, Georg. *The Food and People Dilemma*. North Scituate, Mass.: Duxbury Press, 1973.

Calls for a "reasonable balance" of food and people by coordinating six factors—food production, population control, better storage and utilization of food and feed, nutritional requirements, disease control, and resource appraisal (soil, water, energy).

——. *Too Many: An Ecological Overview of Earth's Limitations*. New York: Macmillan Co., 1971.

Boulding, Kenneth E., and Stahr, Elvis J., eds. *Economics of Pollution*. New York: New York University Press, 1971.

Bradford, Peter. *Fragile Structures: A Story of Oil Refineries, National Security, and the Coast of Maine*. New York: Harper Magazine, 1975.

Brittain, Robert. *Let There Be Bread*. New York: Simon and Schuster, 1952.

Foreword by John Boyd Orr.

Brody, Jane E. "Many Workers Still Face Health Peril Despite Law." *The New York Times*, March 4, 1974, pp. 1, 21.

According to the Public Health Service, one of every ten workers every year suffers a job-related illness or injury, and "prolonged on-the-job exposure to toxic chemicals, dust, noise, heat, cold, radiation and other industrial conditions each year results in the death of at least 100,000 workers and the development of disabling occupational diseases in 390,000 more."

Bronowski, Jacob. *The Ascent of Man.* Boston: Little, Brown & Co., 1974.
Comprehensive and readable history of the harmony of scientific thought and the order of nature.

———. *Science and Human Values.* New York: Julian Messner, 1965.

Brown, Harrison. *The Challenge of Man's Future.* New York: Viking Press, 1954.
An early estimate of global resources. Realistic and more optimistic than many recent American assessments.

———; Bonner, James; and Weir, John. *The Next Hundred Years: A Discussion Prepared for Leaders of American Industry.* New York: Viking Press, 1957.
Future resources as viewed by scientists and industrialists.

———, and Hutchings, Edward, Jr., eds. *Are Our Descendants Doomed? Technological Change and Population Growth.* New York: Viking Press, 1972.

Brown, Lester R. *Seeds of Change: The Green Revolution and Development in the 1970's.* New York: Praeger Publishers, 1970.

Brown, Martin, ed., *The Social Responsibility of the Scientist.* New York: Free Press, 1971.
Scientists on government funding, chemical and biological warfare, food additives, nuclear radiation, disease and social class, ecology, population, land use, food, technology's uses.

Burke, Albert E. "Influence of Man upon Nature—The Russian View: A Case Study." In William L. Thomas, Jr., ed., *Man's Role in Changing the Face of the Earth.* Chicago: University of Chicago Press, 1956, pp. 1035-1051.
Careful examination through the 1940s, by a geographer.

Calder, Nigel, ed. *Nature in the Round: A Guide to Environmental Science.* New York: Viking Press, 1974.
Collection of twenty-six original essays by specialists.

———. *Technopolis: Social Control of the Uses of Science.* New York: Simon and Schuster, 1970.
Role of science in human use of technology.

Caldwell, Lynton Keith. *Environment: A Challenge to Modern Society.* Garden City, N.Y.: Doubleday, 1971.
A political scientist looks at environmental issues from the viewpoint of policy, tasks, management, and action.

———. *In Defense of Earth: International Protection of the Biosphere.* Bloomington: Indiana University Press, 1972.

———; Deville, William B.; and Shuchman, Hedvah L. *Science, Technology, and Public Policy: A Selected and Annotated Bibliography.* Vols. 1 and 2. Bloomington: Indiana University Press, 1969.
5,000 entries.

Carson, Rachel. *Silent Spring.* Boston: Houghton Mifflin Co., 1962, 1973.
Recent landmark in American ecological concern. It may be a death knell

or a distress call which, echoing the cry of the poisoned grebe, will save nature and us.

Castro, Josua de. *The Black Book of Hunger.* New York: Funk and Wagnalls, 1967.

———. *The Geography of Hunger.* Boston: Little, Brown & Co., 1952.

Thesis that improved diet reduces the birth rate.

Cederna, A. *La distruzione della natura in Italia.* Torino: Einaudi, 1975.

Charrier, Jean Bernard. *Où vont villes?* Paris: A. Colin, 1970.

Chemical and Bacteriological (Biological) Weapons and the Effects of Their Possible Use. A United Nations Report. New York: Simon and Schuster, 1970.

"China Explains Her Views on the Population Question." *Peking Review* 16, no. 17 (April 27, 1973): 16-17.

Cobb, John B , Jr., *Is It Too Late? A Theology of Ecology.* Milwaukee: The Bruce Publishing Co., 1971.

A process philosopher addresses important questions.

Cole, LaMont C. "Can the World Be Saved?" *The New York Times Magazine,* March 31, 1968, pp. 35 ff.

———. "The Ecosphere." *Scientific American* 198, no. 4 (April 1958): 83-92.

Collier, John. *Indians of the Americas: The Long Hope.* New York: American Library of World Literature, 1947.

A sympathetic account and scholarly summary of the ecological sense and way of life of the Indians.

———. *On the Gleaming Way.* Chicago: Swallow Press, 1962.

Intimate relation of North American Indians to nature, by a Euro-American who passionately loved his subject.

Committee on Resources and Man, National Academy of Science—National Research Council. *Resources and Man.* San Francisco: W. H. Freeman, 1969.

Committee on Science and Astronautics, U.S. House of Representatives. *Energy Research and Development: Report of the Task Force on Energy of the Subcommittee on Science, Research and Development.* Washington, D.C.: U.S. Government Printing Office, 1973.

Commoner, Barry. *The Closing Circle. Nature, Man and Technology.* New York: Alfred A. Knopf, 1971.

Commoner has been called "the Paul Revere of ecology."

———. *The Poverty of Power. Energy and the Economic Crisis.* New York: Alfred A. Knopf, 1976.

One of the best critics in the United States on the energy question.

———. *Science and Survival.* New York: Viking Press, 1966.

———, et al. "The Integrity of Science." *American Scientist* 1, no. 53 (1965): 174-198.

Costanini, Edmond, and Hanf, Kenneth. "Environmental Concern and Lake Tahoe: A Study of Elite Perceptions, Backgrounds and Attitudes." *Environment and Behavior* 4 (June 1972): 209-242.

A discussion of psychological and social barriers to ecological progress.

Cox, Peter R., and Peel, John, eds. *Population and Pollution.* New York: Academic Press, 1972.

Crenson, Matthew. *The Un-Politics of Air Pollution: A Study of Non-Decision-*

making in the Cities. Baltimore: Johns Hopkins University Press, 1971.

Curtis, Richard, and Hogan, Elizabeth. *Perils of the Peaceful Atom: The Myth of Safe Nuclear Power Plants.* Garden City, N.Y.: Doubleday, Natural History Press, 1969; New York: Ballantine Books, 1970.
   Shows there is no safe way to dispose of radioactive wastes or prevent a nuclear accident, and no economic way to cool the thermal heat produced by nuclear reaction. Since this was written, we have learned from Barry Commoner that the government has suppressed research findings on the advantages of using solar energy.

Daglish, Robert, ed. *The Scientific and Technological Revolution: Social Effects and Prospects.* Moscow: Progress Publishers, 1972.
   Essays from the Soviet Union dealing with the consequences of the scientific-technological revolution in socialist, capitalist, and developing countries.

Dale, E. L., Jr. "The Economics of Pollution." *The New York Times Magazine,* April 19, 1970, pp. 27 ff.

Dansereau, Pierre. *Inscape and Landscape.* New York: Columbia University Press, 1975.

———, ed. *Challenge for Survival: Land, Air, and Water for Man in Megalopolis.* New York: Columbia University Press, 1970.

Darling, F. Fraser. *Wilderness and Plenty.* Boston: Houghton Mifflin Co., 1970.
   Scottish elder scientific statesman of conservation and ecology movements.

———, and Milton, John P., eds. *Future Environments of North America.* Garden City, N.Y.: Doubleday, Natural History Press, 1966.
   This comprehensive, collective volume is a record of a conference convened by the Conservation Foundation in 1965. It explores the organic world and its environment, the development and future of regions, economic patterns and processes, social and cultural purposes, regional planning, and organization and implementation. It is informative, not radical. The tone is expressed by a participant: "If this is what the world is to look like fifty years hence, I am glad that I am almost sixty years old."

Dasmann, Raymond F. *The Conservation Alternative.* New York: John Wiley and Sons, 1975.

———. *Environmental Conservation.* 3d ed. New York: John Wiley and Sons, 1972.
   The author is a leader in applying ecological thought to conservation problems.

———. *Planet in Peril: Man and the Biosphere Today.* New York: UNESCO-World, 1972.

Davidow, Mike. *Cities Without Crisis.* New York: International Publishers, 1976.
   First-hand account of how the Soviet people live—their low-cost housing, food, gas, electricity, telephone, transportation, medical care, sanitoriums, and education. There are problems, but they are of a different kind from those under capitalism.

De Bell, Garrett. *The Environmental Handbook.* Prepared for the First National Environmental Teach-In. New York: Ballantine Books, 1970.

DeNevers, Noel, ed. *Technology and Society.* Reading, Mass.: Addison-Wesley, 1972.

Detwyler, Thomas R., ed. *Man's Impact on the Environment.* New York: McGraw-

Hill Book Co., 1971.
> See Marshall I. Goldman, "Environmental Disruption in the Soviet Union," in this volume for a negative view of Soviet ecology, a view exposed as one-sided by the studies of William M. Mandel.

Development and Environment. The Hague, Paris: Mouton, 1972.
> Working papers and report of the meeting of experts at Founex, Switzerland, June 4-12, 1971, convened by the secretary-general of the United Nations Conference on the Human Environment.

Development of Environmental Protection in Japan. Tokyo: Ministry of Foreign Affairs, n.d.
> Collection of articles published in official sources in the 1970s.

Dice, Lee R. *Man's Nature & Nature's Man.* Westport, Conn.: Greenwood Press, 1973.

DiGiovine, G., and Squillante, R. *Ambiente e potere. L'ecologia e la strategia della partecipazione.* Milano: Etas Libri, 1975.

Disarmament and Development. Report of the Group of Experts on the Economic and Social Consequences of Disarmament. New York: United Nations, 1973.

Disch, Robert, ed. *The Ecological Conscience.* Englewood Cliffs, N.J.: Prentice-Hall, 1970.
> A call for revolution in thought, action, and values.

Dorst, Jean. *Before Nature Dies.* New York: Houghton Mifflin Co., 1970.
> A moving appeal, originally in French.

Douglas, William O. *The Three Hundred Year War: A Chronicle of Ecological Disease.* New York: Random House, 1972.
> Prolific curmudgeon doing battle once more.

——. *A Wilderness Bill of Rights.* Boston: Little, Brown & Co., 1965.

Doxiadis, C. A. *Ekistics: An Introduction to the Science of Human Settlements.* New York: Oxford University Press, 1968.

Dreyfus, Catherine, and Pigeat, Jean Paul. *Les maladies de l'environnement: La France en saccage.* Paris: Denoël, 1971.

Dubos, René. *Man Adapting.* New Haven, Conn.: Yale University Press, 1965.
> Always bringing a wealth of observation and knowledge to his work, Dubos is rather diffuse on social and political matters.

——. *So Human an Animal.* New York: Charles Scribner's Sons, 1968.
> Explorations of how particular human bodies become particular personalities through the formative ecological forces of particular natural and cultural environments.

Duffey, Eric. *Conservation of Nature.* New York: Collins (McGraw-Hill), 1970.
> History of man-nature interactions.

Dumont, René, and Rosier, Bernard. *The Hungry Future.* Trans. from the French by Rosamund Linell and R. B. Sutcliff. New York: Praeger, 1969.
> Criticizes compulsory, premature collectivization of agriculture in socialist nations.

Durrenberger, Robert W. *Environment and Man: A Bibliography.* Palo Alto, Calif.: National Press, 1970.
> 2,200 entries.

Duvigneaud, P., and Tanghe, M. *Ecosystèmes et Biosphère.* Bruxelles, 1967.

Eckholm, Erik P. *Losing Ground: Environmental Stress and World Food Prospects.*
Jointly sponsored by the U.N. Environment Programme and the Worldwatch
Institute, Washington, D.C. New York: W. W. Norton, 1976.

"Ecology and Evolution of Social Organization." *American Zoologist* 14, no. 1
(Winter 1974).
Technical papers on a variety of species. This issue includes also a series of
papers, "Social Play in Mammals."

Economia ed Ecologia. *Atti della XIV riunione scientifica degli economisti.* Milano:
Giuffre, 1975.

Edberg, Rolf. *On the Shred of a Cloud.* Trans. Sven Ahman. University, Ala.: University of Alabama Press, 1969.

Editors of *Fortune. The Environment: A National Mission for the Seventies.* New
York: Harper and Row, 1970.

Ehrenreich, John and Barbara. *The American Health Empire.* New York: Random
House, 1971.

Ehrlich, Paul R. *The Population Bomb.* San Francisco: Sierra Club; New York:
Ballantine Books, 1968.
Millions of young Americans who never heard of Jeremiah have heard of
Ehrlich and his prophecies of doom from hunger and famine in the 1970s.
Perhaps his work has reduced the birth rate.

———, and Ehrlich, Anne H. *Population, Resources, Environment: Issues in Human
Ecology.* 2d ed. San Francisco: W. H. Freeman, 1972.

———, et al. *Human Ecology, Problems and Solutions.* San Francisco: W. H. Freeman,
1973.

Elder, Frederick. *Crisis in Eden: A Religious Study of Man and Environment.*
Nashville, Tenn.: Abingdon Press, 1970.

Ellison, Vivian. *Chemical Invasion: The Body Breakers.* West Haven, Conn.: Pen-
dulum Press, 1972.

Ellul, Jacques. *The Technological Society.* New York: Alfred A. Knopf, 1964.
Pertinent but overstated and pessimistic treatment of technique. Ellul's
dualistic Christianity is presupposed, not argued.

Elton, Charles. *The Ecology of Invasions by Animals and Plants.* New York: John
Wiley and Sons, 1958.
A sensitive and sensible book that argues for conserving the variety of nature because it enriches man and promotes ecological stability. The author's
plea to search "for some wise principle of co-existence between man and
nature" is also a necessary and beautiful prospect for the cultivation of
variety within and among human societies.

Enthoven, Alain C., and Freeman, A. Myrick, 3rd, eds. *Pollution, Resources and
the Environment,* new ed. New York: W. W. Norton, 1974.

Environmental Aspects of Nuclear Power. Vienna: International Atomic Energy
Agency, 1971.

Esposito, John C. *Vanishing Air: The Report on Air Pollution.* New York: Grossman Publishers, 1970.
One of Ralph Nader's Study Group Reports.

Gerasimov, I. P., ed. *Man, Society and the Environment.* Trans. Yuri Shirokov. Moscow: Progress Publishers, 1975.
"Geographical aspects of the uses of natural resources and nature conservation" by a research team of the Institute of Geography of the U.S.S.R. Academy of Sciences. Includes precapitalist and capitalist stages in the man-society-environment interaction, the Soviet experience in developing natural resources and improving the natural environment, natural historical and technical aspects of the problem, and socioeconomic aspects of the problem, such as resource renewal, population and food, urbanization, and recreational facilities.

————; Armand, D. L.; and Yefron, K. M., eds. *Natural Resources of the Soviet Union: Their Use and Renewal.* Trans. Jacek I. Romanovski; English ed., W. A. Jackson. San Francisco: W. H. Freeman, 1971.

Gerecke, Lothar. *How Do We Protect Our Environment? Information, Facts and Data from the GDR.* Berlin: Panorama DDR, 1974.
A succinct, graphic, illustrated account of how an advanced socialist nation, the German Democratic Republic, deals with problems of ecology.

Gill, Don, and Bennett, Penelope. *Nature in the Urban Landscape: A Study of City Ecosystems.* Baltimore: York Press, 1973.

Glacken, Clarence J. *Traces on the Rhodian Shore. Nature and Culture in Western Thought from Ancient Times to the End of the Eighteenth Century.* Berkeley: University of California Press, 1967.
Sui generis history of Western attitudes and activities toward nature.

Glass, D. V., and Eversley, D. E. C., eds. *Population in History: Essays in Historical Demography.* London: Edward Arnold, 1965.

Goldman, Marshall I., ed. *Controlling Pollution: The Economics of a Cleaner America.* Englewood Cliffs, N.J.: Prentice-Hall, 1967.

————. *The Spoils of Progress: Environmental Pollution in the Soviet Union.* Cambridge, Mass.: Harvard University Press, 1972.
There *is* pollution in the Soviet Union, as Goldman contends, but, contrary to his statement, and as William M. Mandel has demonstrated, there is *not* a "convergence of environmental disruption." See William M. Mandel's review in *Environment* 14, no. 10 (December 1972): 43-45, and Goldman's answer plus Mandel's rejoinder in *Environment* 15, no. 4 (May 1973): 29-31.

Goldsmith, Edward, and Editors of the *Ecologist. Blueprint for Survival.* Boston: Houghton Mifflin Co., 1972.

Goldsmith, Maurice, and Mackay, Alan, eds. *Society and Science.* New York: Simon and Schuster, 1965.
Festschrift honoring J. D. Bernal's *The Social Function of Science.* The contributors are J. D. Bernal, P. M. S. Blackett, E. H. S. Burhop, Herbert Coblans, J. B. S. Haldane, Peter Kapitsa, Alexander King, Maurice Korach, Joseph Needham, Gerard Piel, N. W. Pirie, C. F. Powell, C. P. Snow, D. J. de Solla Price, and R. L. M. Synge.

Goldsmith, Walter, ed. *Worlds of Man: Studies in Cultural Ecology.* Chicago: Aldine Publishing Co., 1972.

Graham, Frank, Jr. *Since Silent Spring.* Boston: Houghton Mifflin Co., 1970.

Graham, Loren R. *Science and Philosophy in the Soviet Union*. New York: Alfred A. Knopf, 1972.

The first of its kind, in scope and content, by an American scholar—after fifty-five years of history of the Soviet Union. Detailed and thorough.

Guthrie, Daniel A. "Primitive Man's Relationship to Nature." *Bioscience* 21 (July 1, 1971): 721-723.

It was not without adverse affects.

Hall, Edward T. *The Hidden Dimension*. Garden City, N.Y.: Doubleday, 1969.

Distance as a function of relations among animals and men, four kinds of relations, cultural variations, and implications for living conditions.

Hall, Gus. *Ecology: Can We Survive Under Capitalism?* New York: International Publishers, 1972.

American Communist's critique and corrective.

———. *The Energy Rip-Off: Cause and Cure*. New York: International Publishers, 1974.

On U.S. and global monopoly control of the petroleum industry and the way out through national liberation, nationalization, and socialism.

Halprin, Lawrence. *The RSVP Cycles: Creative Processes in the Human Environment*. New York: George Braziller, 1970.

Hamilton, Michael, ed. *This Little Planet*. New York: Charles Scribner's Sons, 1970.

Hammond, Allen L.; Metz, William; and Maugh, Thomas H., II. *Energy and the Future*. Washington, D.C.: American Association for the Advancement of Science, 1973.

Handler, Philip, ed. *Biology and the Future of Man*. New York: Oxford University Press, 1970.

Described by C. H. Waddington as "by far the best single volume of the whole biology that is now available anywhere."

Hardin, Clifford M., ed. *Overcoming World Hunger*. Englewood Cliffs, N.J.: Prentice-Hall, 1969.

Hardin, Garrett. *Exploring New Ethics for Survival. The Voyage of the Spaceship Beagle*. New York: Viking Press, 1972.

An eminent scientist's literary and probing work on progress, pollution, and population. See the essay, "The Tragedy of the Commons," which calls for "mutual coercion, mutually agreed upon," to secure population control. Superficial dismissal of Marxism.

———. *Population, Evolution, and Birth Control*. 2d ed. San Francisco: W. H. Freeman, 1969.

Harte, John, and Socolow, Robert H. *Patient Earth*. New York: Holt, Rinehart and Winston, 1971.

Hayashi, Yujiro, ed. *Perspectives on Postindustrial Society*. Tokyo: University of Tokyo Press, 1970.

Hayter, R., and Cox, J. "The Crisis of Man and Environment." *Marxism Today* 15, no. 9 (September 1971): 261-269.

Heilbroner, Robert, *Between Capitalism and Socialism*. New York: Random House, 1970.

Helfrich, Harold W., ed. *The Environmental Crisis*. New Haven, Conn.: Yale University Press, 1970.

Essays by authorities from a variety of fields.

Hemming, James. *Probe 2— Values for Suvival.* London: Angus and Robertson, 1974.

Henderson, Lawrence J. *The Fitness of the Environment: An Inquiry into the Biological Significance of the Properties of Matter.* New York: Macmillan Co., 1913; Boston: Beacon Press, 1958.

An early classic, linking life and matter, organism and environment.

Henkin, Haromon, et al., eds. *The Environment, the Establishment, and the Law.* Boston: Houghton Mifflin Co., 1971.

Henry, S. Mark, ed. *Symbiosis: Its Physiological and Biochemical Significance.* New York: Academic Press, 1966.

Higbee, Edward C. *The Squeeze: Cities Without Space.* London: William Morrow, 1961.

Hinton, Alice Mary, ed. *Against Pollution and Hunger.* Oslo: Universitetsforlaget; New York: Halsted Press, 1974.

Hockett, Charles Francis. *Man's Place in Nature.* New York: McGraw-Hill Book Co., 1973.

The author's theme is: "We are part of Earth, and our history is part of its history." Strong on anthropological origins, weak on political economy.

Holdren, John P., and Ehrlich, Paul R., eds. *Global Ecology: Readings Toward a Rational Strategy for Man.* New York: Harcourt, Brace, Jovanovich, 1971.

Holdren, John, and Herrera, Philip. *Energy: A Crisis in Power.* San Francisco and New York: Sierra Club Books, 1971.

Hoshino, Y. *Pollution of the Inlands Sea (Setonaikai) of Japan.* Tokyo: Iwanami, 1972.

(In Japanese.)

Hsin, Fang. "Economic Development and Environmental Protection." *Peking Review* 16, no. 29 (July 20, 1973): 6-8.

Hughes, J. Donald. *Ecology and Ancient Civilization.* Albuquerque: University of New Mexico Press, 1975.

Hutchinson, G. Evelyn. *The Ecological Theater and the Evolutionary Play.* New Haven, Conn.: Yale University Press, 1965.

Huth, Hans. *Nature and the American: Three Centuries of Changing Attitudes.* Berkeley: University of California Press, 1957; Lincoln: University of Nebraska Press, 1972.

"The basic developments which led to the conservation movement in this country." Fascinating and on the whole optimistic history of our cultural-natural heritage.

Huxley, Aldous. *The Politics of Ecology.* Santa Barbara, Calif.: Center for the Study of Democratic Institutions, 1963.

Huxtable, Ada Louise. *Will They Ever Finish Bruckner Boulevard?* New York: Macmillan Co., 1970.

Her articles from *The New York Times* on urban environment.

Iltis, Hugh H.; Loucks, Orie L.; and Andrews, Peter. "Criteria for an Optimum Human Environment." *Bulletin of the Atomic Scientists* 26, no. 1 (January 1970): 2-6.

Poses and explores research issues in the study of genetic-evolutionary human needs and healthy human-environmental relations.

International Conference on Family Planning Programs. *Family Planning and Population Programs: A Review of World Developments.* Proceedings of Conference in Geneva, August, 1965. Chicago: University of Chicago Press, 1966.

International Food Policy Research Institute. *Meeting Food Needs in the Developing World: The Location and Magnitude of the Task in the Next Decade.* February 1976.

Jackson, Wes, ed. *Man and the Environment.* Dubuque, Iowa: William C. Brown, 1971.

Collection of a variety of short excerpts, many from *Science,* with emphasis on problems of population-food and environmental destruction.

Jacobs, Jane. *The Death and Life of Great American Cities.* New York: Random House, 1961.

———. *The Economy of Cities.* New York: Random House, 1969.

Jacobson, Nolan Pliny. *Buddhism: The Religion of Analysis.* Carbondale: Southern Illinois University, 1970.

This book opens up the richness of Buddhist thought regarding the qualitative flow and togetherness of nature's events. The Buddhists arrived at an ecological perspective intuitively, without knowing science.

Jacoby, Neil H., and Pennar, F. G. *The Polluters: Industry or Government.* Levittown, N.Y.: Transatlantic Arts, 1972.

Jarrett, Henry, ed. *Environmental Quality in a Growing Economy.* Baltimore: Johns Hopkins University Press, 1966.

Jessop, N. M. *Biosphere: A Study of Life.* Englewood Cliffs, N.J.: Prentice-Hall, 1970.

Johnson, Cecil E., ed. *Ecocrisis.* New York: John Wiley and Sons, 1970.

Sixteen articles on pollution, population, war, the environment, and the future.

Johnson, E. A. J. *The Organization of Space in the Developing Countries.* Cambridge, Mass.: Harvard University Press, 1971.

Johnson, Huey D., ed. *No Deposit—No Return. Man and His Environment: A View Toward Survival.* Reading, Mass.: Addison-Wesley, 1970.

Johnson, Stanley P. *The Politics of the Environment: The British Experience.* North Pomfret, Vt.: David and Charles, n.d.

Jouvenel, Bertrand de. *The Art of Conjecture.* Trans. Nikita Lary. New York: Basic Books, 1967.

Need for forecast of futures.

Jungk, Robert, and Galtung, Johan, eds. *Mankind 2000.* London: Allen and Unwin, 1969.

Publication of the International Peace Research Institute in Oslo, Norway.

Kaino, M., ed. *Studies About Law on Kogai.* Tokyo: Nihonhyoron-sha, 1969.

(In Japanese.)

Kogai means pollution.

Kalyadin, Alexander, and Kade, Gerhard, eds., *Detente and Disarmament: Problems and Perspectives.* Vienna: Gazzetta Publishing House, 1976.

Published by the International Institute for Peace, this book contains up-to-

date articles by specialists and documents from 1975-1976 forums on disarmament.

Kapitza, Peter. "Basic Factors in the Organization of Science and How They Are
Handled in the U.S.S.R." *Daedalus* 102 (Spring 1973): 167-176.

Kapp, K. William. *The Social Costs of Private Enterprise*. New York: Shocken Books,
1971.
Critique of the theory and practice of entrepreneurial investment and profit
and the consequent impairment and destruction of the physical and *human*
environment for ourselves and our posterity. First published in 1950.

Kay, David A., and Skolnikoff, Eugene B., eds. *World Eco-Crisis: International Organizations in Response*. Madison: University of Wisconsin Press, 1972.

Keldysh, M. V.; Millionshchikov, M. D.; and Fedoseev, P. N., eds. *Science in the
USSR: To the 50th Anniversary of the Formation of the Union of Soviet
Socialist Republics 1922-1972*. Moscow: Progress Publishers, 1972.
Essays by academicians on science in each of the republics.

Kenya's National Report to the United Nations on the Human Environment.
Nairobi: National Committee on the Human Environment, 1972.

Kepes, Gyrogy, ed. *Arts of the Environment*. New York: George Braziller, 1972.

Kimball, Thomas L. "I Felt the Winds of Change in Russia." *Natural Wildlife* 11
(February-March 1973): 4-11.
U.S.-U.S.S.R. negotiations on environmental cooperation.

Klausner, Samuel. *On Man in His Environment*. San Francisco: Jossey-Bass, 1971.

Kozlovsky, Daniel G., ed. *An Ecological and Evolutionary Ethic*. Englewood Cliffs,
N.J.: Prentice-Hall, 1974.

Krutch, Joseph W. *The Best Nature Writing of Joseph Wood Krutch*. New York:
William Morrow, 1970; Simon and Schuster Pocket Books, 1971.
Rambles through nature and the literature about nature.

Kuczynski, Jurgen, and Nicolson, B. "Lenin and the Energy Question." *Labour
Monthly* (March 1974): 129-132.

Kuenzlen, Martin. *Playing Urban Games: The Systems Approach to Planning*. New
York: George Braziller, 1972.
Critique of the inhumanism of city planners.

LaBarre, Weston. *The Human Animal*. Chicago: University of Chicago Press, 1955.
The evolution of the ecology of the human family (adult female, adult male,
and child).

Lacey, Michael J. "Man, Nature, and the Ecological Perspective." *American Studies.
An International Newsletter* 8, nos. 1-3 (Spring 1970): 13-27.
Valuable history and references.

Landers, Richard R. *Man's Place in the Dybosphere*. Englewood Cliffs, N.J.: Prentice-
Hall, 1966.
Man in a future world of machines.

Landsberg, Hans H.; Fischman, Leonard L.; and Fisher, Joseph L. *Resources in
America's Future: Patterns of Requirements and Availabilities 1960-2000*.
Baltimore: Johns Hopkins University Press, 1963.

Lapp, Ralph E. *The Logarithmic Century*. Englewood Cliffs, N.J.: Prentice-Hall,
1973.

Larner, Jeremy, and Howe, Irving, eds. *Poverty: Views from the Left*. New York: William Morrow, 1971.

Lefebvre, Henri. *Le droit à la ville*. Paris: Anthropos, 1968.

Leiss, William. *The Domination of Nature*. New York: George Braziller, 1972.

Lenin, V. I. *The Development of Capitalism in Russia: The Process of the Formation of a Home Market for Large-Scale Industry*. Moscow: Foreign Languages Publishing House, 1956.

Lenin's description (ch. 6, section 7) of the dehumanization of the worker in capitalist domestic industry in Russia parallels Marx's and Engels' description of English workers.

——. *Materialism and Empirio-criticism*. In *Collected Works*, vol. 14. Moscow: Progress Publishers, 1968.

Criticism of idealist Machians and defense of materialism based on natural sciences. This is the main repository of Lenin's views on nature and man's relation to it, coupled with his ideas about dialectics (see his *Philosophical Notebooks*, vol. 38, in *Collected Works*, edition cited above).

——. "Our Foreign and Domestic Position and the Tasks of the Party." Speech of November 21, 1920, in *Collected Works*, vol. 31. Moscow: Progress Publishers, 1966, pp. 408-426.

Lenin gives his formula for man's social success through political democracy and the tapping of nature's energies: "Communism is Soviet power plus the electrification of the whole country, since industry cannot be developed without electrification."

——. "Report on the Work of the Council of People's Commissars to the Eighth All-Russia Congress of Soviets," December 22, 1920. In *Collected Works*, vol. 31. Moscow: Progress Publishers, 1966, pp. 487-518.

Lenin repeats his definition of communism and points out that it is not electricity that is "unnatural" (contrary to a peasant) but "backwardness, poverty and oppression under the yoke of the landowners and the capitalists."

Lentz, Theo. F. *Towards a Technology of Peace*. St. Louis: Peace Research Laboratory, 1972.

Groundbreaking work by the author of *Toward a Science of Peace* and the founder and director of the Peace Research Laboratory in St. Louis.

Leopold, Aldo. *A Sand County Almanac with Essays on Conservation from Round River*. New York: Oxford University Press, 1949, 1953; Ballantine Books, 1970.

Criticizes the trophy hunting of "man the conqueror" and favors the perception and receptivity of a "land ethic" and a "conservation esthetic." Leopold only mentions in passing the principal enemy of such a noble ethics, namely, "profit."

Linton, Ron M. *Terracide: America's Destruction of Her Living Environment*. Boston: Little, Brown, 1970.

Livingston, John A. *One Cosmic Instant: Man's Fleeting Supremacy*. New York: Houghton Mifflin Co., 1973.

Longgood, William. *The Darkening Land*. New York: Simon and Schuster, 1972.

"Attempts to explain the world's intricate interrelationships and how pollutants interfere with the life-support systems."

Loth, David, and Ernst, Morris L. *The Taming of Technology.* New York: Simon
and Schuster, 1972.
Discusses legal control of pollution, food additives, organ transplants, and
invasions of privacy.

Lovins, Amory. *Non-Nuclear Futures: The Case for an Ethical Energy Strategy.*
Philadelphia: Ballinger Publishing Co., 1975.

——. *The Stockholm Conference: Only One Earth.* San Francisco: Friends of the
Earth, 1972.

——. *World Energy Strategies.* Philadelphia: Ballinger Publishing Co., 1975.

Luce, Gay Gaer. *Body Time: Physiological Rhythms and Social Stress.* New York:
Pantheon Books, 1971.

Maddox, John. *The Doomsday Syndrome.* New York: McGraw-Hill Book Co., 1972.
Sanguine and critical of Carson, Ehrlich, Dubos, etc.

Maison des Sciences de l'Homme Symposium. *Political Economy of Environment;
Problems of Method.* New York: 1 Humanities Press, 1972.

"Man and His Environment." *Soviet Studies in Philosophy* 13, no. 2-3 (Fall-Winter
1974-1975).
A series of round table discussions on the philosophical problems of ecology,
translated into English from *Voprosy filosofii,* nos. 1-4 (1973).

"Man and Nature." Editorial in *Soviet Life* 119 (August 1966): 3.
Cooperation and dynamic balance are emphasized.

Man Conquers Nature. London: Society for Cultural Relations with the U.S.S.R.,
1952.
Scientists, such as S. M. Manton, J. D. Bernal, and F. LeGros Clark, on new
Soviet construction schemes. Socialist enthusiasm.

Man in the Living Environment. A Report on Global Ecological Problems. Published
for the Institute of Ecology. Madison: University of Wisconsin Press, 1972.

*Man—Science—Technology. A Marxist Analysis of the Scientific-Technological
Revolution.* Moscow-Prague: Academia Prague, 1973.
Contains works by members of the three organizers of the volume—the Institute of Philosophy of the Academy of Sciences of the U.S.S.R., the Institute of Philosophy and Sociology of the Czechoslovak Academy of Sciences, and the Institute of the History of National Sciences and Technology
of the Academy of Sciences of the U.S.S.R. "Attempts an analysis of the
interconnected phenomena and processes occurring in the world under the
impact of today's scientific and technological revolution."

Mandel, William M. "The Soviet Ecology Movement." *Science and Society* 36, no.
4 (Winter 1972): 385-416.

Margalef, Ramon. *Perspectives in Ecological Theory.* Chicago: University of Chicago
Press, 1968.

Martinez, G. and Mautino, F. *Tutela del patrimenio storico, artistico, naturale, e
disciplina urbanistica.* Firenze: Le Monnier, 1969.

Marx, Leo. *The Machine in the Garden: Technology and the Pastoral Ideal.* New
York: Oxford University Press, 1964.

Contrast between the eighteenth-century dream of North American immigrants and the nineteenth-century spirit of conquest and production of material wealth—a conflict that still haunts us in the United States.

Marx, Wesley. *The Frail Ocean.* New York: Coward-McCann, 1967.

Matley, I. A. "The Marxist Approach to the Geographical Environment." *Annals of the Association of American Geographers* 56 (1966): 97-111.

Matthews, William H.; Smith, Frederick E.; and Goldberg, Edward D., eds. *Man's Impact on Terrestrial and Oceanic Ecosystems.* Cambridge, Mass.: M.I.T. Press, 1971.

Mbiti, John S. *African Religions and Philosophies.* New York: Praeger Publishers, 1969.
The unity of the spiritual world and the physical universe among African peoples.

McHale, John. *The Ecological Context: Energy and Materials.* Carbondale: Southern Illinois University. World Resources Inventory. World Design Science Decade 1965-1975. Document 6, 1967.

————. *The Future of the Future.* New York: George Braziller, 1969.
Future studies, the need for a global society.

McHarg, Ian. *Design with Nature.* Garden City, N.Y.: Doubleday, 1971.
Broad knowledge, incisive criticism, aesthetic sensitivity, and poetic passion—lacking only in class analysis.

McHenry, Robert, and Van Doren, Charles, eds. *A Documentary History of Conservation in America.* New York: Praeger Publishers, 1972.

Meadows, Dennis L. *Toward Global Equilibrium.* New York: John Wiley and Sons, 1973.

————, et al. *Dynamics of Growth in a Finite World.* Cambridge, Mass.: Wright-Allen, 1974.

Meadows, Donella H.; Meadows, Dennis L; Randers, Jørgen; and Behrens, William W., III. *The Limits to Growth. A Report for the Club of Rome's Project on the Predicament of Mankind.* New York: Universe Books, 1972.
Because the demands of the human population in the form of food production, land use, industrial processing of nonrenewable resources, energy production, and waste disposal are increasing exponentially, and in most countries the population growth rate exceeds the economic growth rate, the authors propose "self-imposed limitation to growth." Their pessimistic projections have been modified in subsequent work.

Meek, Ronald L., ed. *Marx and Engels on the Population Bomb.* Palo Alto, Calif.: Ramparts Press, 1971.

Meighan, Don G. "How Safe Is Safe Enough?" *The New York Times Magazine,* June 20, 1976, pp. 8 ff.
"For the time being, perhaps until solar and fusion systems are brought to a state of readiness, it would seem that the only energy source available to the nation is nuclear fission, warts and all."

Melman, Seymour. *Our Depleted Society.* New York: Dell Publishing Co. 1965.
Extensive factual indictment—"an economic audit of the price that America has paid for twenty years of the cold war."

Mesarovic, Mihailo, and Pestel, Eduard. *Mankind at the Turning Point. The Second Report to the Club of Rome.* New York: E. P. Dutton, 1974.

Meshenberg, Michael. *Environmental Planning: A Selected Annotated Bibliography.* Chicago: American Society of Planning Officials, 1970.

Mesthene, Emmanuel G. *Technological Change: Its Impact on Man and Society.* Harvard Studies in Technology and Society. Cambridge, Mass.: Harvard University Press, 1970; New York: New American Library, 1970.

Meyer, Alfred, ed. *Encountering the Environment.* London: Van Nostrand, 1971. Essays from articles published in *Natural History.*

Meyer, Jon K. *Bibliography on the Urban Crisis: The Behavioral, Psychological, and Sociological Aspects of the Urban Crisis.* Publication no. 1948. Washington, D.C.: U.S. Public Health Service, 1969.

Miller, Herman P. *Rich Man, Poor Man.* New York: Thomas Y. Crowell, 1971. Affluence and economic inequality as causes of many problems.

Miller, R. S.; Woodwell, G. M.; Burch, W. R.; Jordan, P. A.; and Means, R. L. *Man and His Environment: The Ecological Limits of Optimism.* Yale University School of Forestry Bulletin No. 76. New Haven, Conn.: Yale University Press, 1970.

Mishan, Ezra J. *The Costs of Economic Growth.* New York: Praeger Publishers, 1967.
Conservative but hardheaded.

Mitchell, John G., and Stallings, Constance L. eds. *Ecotactics: The Sierra Club Handbook for Environmental Activists.* New York: Simon and Schuster Pocket Books, 1970.

Miyamoto, Ken-ichi. "Japanese Capitalism at a Turning Point." *Monthly Review* 26, no. 7 (December 1974): 15-29.

———. "Japan's Postwar Economy and Pollution Problems." *Business Review* (Osaka City University) 26, no. 2 (July 1975): 95-125.

———. *Kogai vs. People's Movement.* Tokyo: Jichitai-kenkyu-sha, 1970. (In Japanese.)

Montagu, Ashley. *The Direction of Human Development.* Rev. ed. New York: Hawthorn Books, 1970.
Drawing on a wide range of data, the author constructs a theory of human development based on needs, especially love. Of particular value are the chapter and bibliography on "the biological basis of cooperation."

Moore, John A., comp. *Science for Society: A Bibliography.* Washington, D.C.: Commission on Science Education, American Association for the Advancement of Science, 1970.

Moore, Wilbert E., ed. *Technology and Social Change.* Chicago: Quadrangle Books, 1972.
Articles from the last twenty-five years of *The New York Times Magazine.*

Morison, Elting E. *From Know-How to Nowhere.* New York: Basic Books, 1975.

Morse, Elsa Peters. *The Key to World Peace and Plenty.* San Francisco: 1 Summit, 1960.

Muller, C., ed. *Ecology and Mental Health.* White Plains, N.Y.: Albert J. Phiebig, 1975.

Mumford, Lewis. *The City in History: Its Origins, Its Transformations, and Its Prospects.* New York: Harcourt, Brace and World, 1961.

——. *The Culture of Cities.* New York: Harcourt, Brace, Jovanovich, 1970.

——. *The Pentagon of Power: The Myth of the Machine.* Vol. 2. New York: Harcourt, Brace, Jovanovich, 1970.

Devastating melancholic meditations on "the wholesale miscarriages of megatechnics." An eminent moralist in our time, Mumford here is long on humanistic insight, short on class analysis.

——. *Technics and Civilization.* New York: Harcourt, Brace, 1934.

Fertile technical history of the last 1000 years on the "reciprocal and many-sided" relation between technics and other domains of culture. Assesses human values of machines (e.g., objectivity) and their perversions. Calls for "assimilation of the machine" into the organic world of person and community, and excoriates "the vicious paradox of capitalist production." Written during the Great Depression, capitalism's most severe crisis to date, this work advocates "basis communism" (confusedly described as "post-Marxian") and is Mumford's most radical work.

——. *The Urban Prospect.* New York: Harcourt, Brace and World, 1968.

Emphasis on planned and integral communities.

Murphey, Rhoads. "Man and Nature in China." *Modern Asian Studies* 1, no. 4 (1976): 313-333.

Historical and present-day attitudes.

Murphy, Earl F. *Man and His Environment: Law.* Man and His Environment Series. New York: Harper and Row, 1971.

Myrdal, Alva. *The Game of Disarmament: How the United States and the Russians Run the Arms Game.* New York: Pantheon Books, 1977.

Myrdal, Gunnar. *The Challenge of World Poverty.* New York: Pantheon Books, 1970.

Nader, Ralph. "The Scientist and His Indentured Professional Societies." *Bulletin of Atomic Scientists* 28, no. 2 (February 1972): 43ff.

Scientific knowledge should be made public.

——, et al. *The Environment Committees. A Study of the House and Senate Interior, Agriculture and Science Committees.* ed. Peter H. Schuck. New York: Grossman Publishers, 1975.

Nash, Roderick. *Wilderness and the American Mind.* New Haven, Conn.: Yale University Press, 1967.

Wilderness as both factual condition and romantic symbol in the context of American civilization.

National Academy of Sciences. *Symposium on Aid and Threats to Society from Technology.* Washington, D.C., 1970.

——, Committee on Science and Astronautics. *Technology: Processes of Assessment and Choice.* Washington, D.C., 1969.

—— and National Research Council. *Resources and Man: A Study and Recommendatio by the Committee on Resources and Man.* San Francisco: W. H. Freeman, 1969.

"Nature: A Shrine or a Workshop?" *Sputnik,* October 1973, pp. 41-57.

A Soviet view.

Nature Protection in the Soviet Union. From the Record of the Fourth Session of the U.S.S.R. Supreme Soviet, September 1972. Moscow: Novosti, 1974.

Needham, Joseph. *Human Law and the Laws of Nature in China and the West.* London: Oxford University Press, 1951.

———. *Science and Civilisation in China.* Vols. 1-5. Cambridge, Mass., and New York: Cambridge University Press, 1954-1975.
    All-time classic that has just about everything as a scholarly study.

Neuhaus, Richard. *In Defense of People.* New York: Macmillan Co., 1971.
    Criticism of environmentalists who overlook the poor.

Nicholson, Max. *The Environmental Revolution.* New York: McGraw-Hill Book Co., 1970.

Nicol, Hugh. *The Limits of Man: An Enquiry into the Scientific Bases of Human Population.* London: Constable, 1967.

Novik, I. B., et al., eds. *Methodological Aspects of the Study of the Biosphere.* Moscow: Nauka, 1975. (In Russian.)

Novikov, G. A., et al., eds. *Essays on the History of Ecology.* Moscow: Nauka, 1970. (In Russian.)

Odum, Eugene P. *Fundamentals of Ecology.* 3d ed. Philadelphia: W. B. Saunders, 1971.
    The textbook most used in college courses in ecology.

———, et al. *The Crisis of Survival.* Glenview, Ill.: Scott, Foresman, 1970.

Odum, Howard T. *Environment, Power and Society.* New York. Wiley-Interscience, 1971.
    Man's survival problems are attacked with the tools of energy principles, ecological thought, and systems analysis.

The Onyx Group, Inc., comp. *Environment U.S.A.: A Guide to Agencies, People, and Resources.* Ann Arbor, Mich.: R. R. Bowker Co., 1974.
    Compilation of information about agencies, organizations, consulting firms, individuals, environmental officers of corporations, unions, educational programs, libraries, conferences and meetings, films, newspapers, radio and TV stations, environmental law, fund raising, and employment; with bibliography, glossary, and indexes.

Opie, John, ed. *Americans and Environment: The Controversy Over Ecology.* Indianapolis, Ind.: D. C. Heath, 1971.

Osborn, Fairfield. *Our Plundered Planet.* Boston: Little, Brown & Co., 1948; New York: Pyramid, 1968.
    Once a voice in the wilderness, in retrospect a sounder of a worldwide theme. "Man's conflict with nature" and the necessity of comprehending and cooperating with nature.

———. *The Limits of the Earth.* Boston: Little, Brown & Co., 1953.

Owings, Nathaniel A. *The American Aesthetic.* New York: Harper and Row, 1969.
    Illustrated and analyzed ugliness of American cities and countryside; emphasizes a need for new values.

Packard, Vance. *The Waste Makers.* New York: David McKay, 1960.
    New ways of disposing of the surplus-product.

Paddock, William, and Paddock, Paul. *Famine—1975!* Boston: Little, Brown & Co., 1967.
> Prediction of the end of surplus food production in North America between 1975 and 1980, termination of food shipments to underdeveloped countries, and consequent mass starvation. Exploration of emergency measures—land use, fertilizers, pesticides, and nuclear war.

Parker, John. *Discovery: Developing Views of the Earth from Ancient Times to the Voyages of Captain Cook.* New York: Charles Scribner's Sons, 1972.

Parsons, Howard L. *Man East and West: Essays in East-West Philosophy.* Amsterdam: B. R. Grüner, 1975.
> Man's destructive and constructive relations with his fellow man and nature as revealed in philosophical and religious literature.

———. *Man Today—Problems, Values, and Fulfillment.* Vols. 4/5, *Revolutionary World.* Amsterdam: B. R. Grüner, 1974.
> Chapters on war, American society, technology, values, peaceful coexistence, evolution, ecology, education, space, wonder.

———. *Self, Global Issues, and Ethics.* Vols. 21/22, *Revolutionary World.* Amsterdam: B. R. Grüner, 1977.
> Includes essays on multinational corporations, neo-Malthusianism and socialism, peaceful coexistence and national liberation, detente, disarmament, development, and global ethics.

Passmore, John. *Man's Responsibility for Nature.* New York: Charles Scribner's Sons, 1974.

Peaceful Uses of Atomic Energy. 15 vols. New York: Unipub, 1972.

Pecsi, M., and Probald, F., eds. *Man and His Environment.* Budapest: Akademiai Kiado, 1974.

Peixoto, Jose P., and Kettani, M. Ali. "The Control of the Water Cycle." *Scientific American* 228, no. 4 (April 1973): 46-61.

People's Sovereignty over Raw Materials: Essence of Development. Helsinki: World Peace Council, 1974.
> Pamphlet that links economic detente to political detente, development to peace. Includes the important Declaration of the UN General Assembly on the Establishment of a New International Economic Order (1974).

Perlo, Victor. "Over-Reacting over Reactors." *Daily World,* July 11, 1974, pp. M2-M3.
> A leading U.S. Marxist economist, using the Soviet experience as background, sees nuclear power plants as the best present alternative to the developing of solar energy. He criticizes the opponents of nuclear power plants for failing to struggle against atomic weapons and to work toward nationalization of fuel and power industries.

Perloff, Harvey S., ed. *The Quality of the Urban Environment: Essays on "New Resources" in an Urban Age.* Baltimore: Johns Hopkins University Press for Resources for the Future, 1969.

Phillipson, John. *Ecological Energetics.* New York: St. Martin's Press, 1966.

Podniesinski, Antoni. "Environmental Protection in Poland." *New Perspectives* 6, no. 3 (1976): 24.

Political Economy of Environment. Problems of Method. Paris: Ecole Pratique des Hautes Etudes; The Hague, Paris: Mouton, 1972.

Potter, Van Rensselaer. *Bioethics: Bridge to the Future.* Englewood Cliffs, N.J.: Prentice-Hall, 1971.

Problems of the Human Environment in Japan. Report to the United Nations, March 31, 1971. Japan Reference Series No. 2-71. Tokyo: Public Information Bureau, Ministry of Foreign Affairs, 1971.

"Protecting the Environment." *World Marxist Review* 15, no. 6 (June 1972): 16-49.
    This is a report of a discussion seminar in Prague, March 29-31, 1972, attended by Marxist scientists and representatives of Communist and Workers' parties from thirty-six countries.

"Protection of the Environment—Protection of Man," and "Safeguarding Nature." *Hungarian Review* 10 (1976): 2-5.

Pryde, Philip. *Conservation in the Soviet Union.* Cambridge, Mass.: Cambridge University Press, 1972.
    Includes favorable account of Soviet environmental education in the schools and extracurricular activities.

Pursell, Carroll, ed. *From Conservation to Ecology: The Development of Environmental Concern.* New York: T. Y. Crowell, 1973.

Quality of the Environment in Japan. Environmental Agency, Japan, 1975.
    "A condensed version of the White Paper on the Environment submitted to the 73rd Session of the Diet in pursuance of the Basic Law for Environmental Pollution Control."

Rathlesberger, James, ed. *Nixon and the Environment: The Politics of Devastation.* New York: Village Voice, 1972.

Rattansi, P. M., et al. *Science and Society: 1600-1900.* New York: Cambridge University Press, 1972.
    Six essays on specific periods in the history of science.

Redfield, Robert, ed. *Levels of Integration in Biological and Social Systems. Biological Symposia,* vol. 8. Lancaster, Pa.: Jacques Cattell, 1942.
    Biologists and social scientists of the University of Chicago deal with the question, "How are parts constituted into wholes, throughout the range of life forms?"

Reich, Charles A. *The Greening of America.* New York: Random House, 1970.
    Call for a new "consciousness" as the way to transcend the country's material and cultural hangups. American idealism *in extremis*.

Reissman, Leonard. *The Urban Process: Cities in Industrial Societies.* New York: Free Press, 1964.

Report of the Governing Council of the United Nations Environment Programme on the Work of Its Fourth Session. Nairobi: United Nations Environment Programme, 1976.
    This session took place March 30 to April 14, 1976, in Nairobi. The UNEP

was established in 1972 by the General Assembly of the United Nations to implement the broad cooperation on environmental problems recommended at the first U.N. Conference on the Human Environment, held at Stockholm in June 1972.

Report of the United Nations Conference on the Human Environment. Stockholm, June 5-16, 1972. New York: United Nations, 1973.

Report of the United Nations Scientific Committee on the Effects of Atomic Radiation. Official Record of the General Assembly, 13th Session, Supplement No. 17, 1958; 17th Session, Supplement No. 16, 1962; 19th Session, Supplement No. 14, 1964.

Report of the World Food Conference. Rome, November 5-16, 1974. New York: United Nations, 1975.

"Report on the Stockholm Conference. The UN Conference on the Human Environment at Stockholm, June, 1972." *Bulletin of Atomic Scientists* 28, no. 9 (September 1972): 16-56.

Resources of the Biosphere on the Territory of the USSR. Scientific Foundations of Their Rational Utilisation and Protection. Moscow, 1971.
A Soviet national report prepared for the UNESCO conference on the biosphere in Paris in 1968.

Revelle, Roger, et al., eds. *The Survival Equation: Man, Resources, and His Environment.* Boston: Houghton Mifflin Co., 1971.

———, and Landsberg, Hans L., eds. *America's Changing Environment.* Boston: Houghton Mifflin Co., 1970.
Based on the Fall 1967 issue of *Daedalus.*

Ridgeway, James. *The Politics of Ecology.* New York: E. P. Dutton, 1971.
Leftist attack on establishment policy.

Roelofs, Robert T.; Crowley, Joseph N.; and Hardesty, Donald L., eds. *Environment and Society.* Englewood Cliffs, N.J.: Prentice-Hall, 1974.
Readings and bibliographies on man-nature themes, technology and progress and the environment, survival and the quality of life, the environment and the new economics, political reconstruction, prospects for the city, global perspectives, and an environmental ethic.

Rose, David J. "Energy Policy in the U. S." *Scientific American* 230, no. 1 (January 1974): 20-29.
"Recent steps to alleviate energy shortages leave the long-range problem unsolved."

Roszak, Theodore. *The Making of a Counter Culture.* Garden City, N.Y.: Doubleday, 1969.
A guru of the disaffected young. Antiscience, antitechnology, antitechnocracy.

Rothenblatt, Ben, ed. *Changing Perspectives on Man.* Chicago: University of Chicago Press, 1968.

Rudd, Robert L. *Pesticides and the Living Landscape.* Madison: University of Wisconsin Press, 1964.

Ryan, William. *Blaming the Victim.* New York: Pantheon Books, 1971.
Middle-class scapegoating of the poor.

Sax, Joseph L. *Defending the Environment.* New York: Alfred A. Knopf, 1971.

Schmidt, Alfred. *The Concept of Nature in Marx.* New York: Humanities Press, 1972. .
> Detailed study of the sources. Erroneously ascribes to Engels the view that man is "a passive reflection of the process of nature" and describes Marx as a great "pessimist."

Schrag, Peter. "Who Owns the Environment?" *Saturday Review* 52, no. 27 (July 4, 1970): 6-9 ff.
> Inconclusive—but "the most likely course of action" is "to reduce the incentive to amass private wealth and to provide adequate social services."

Schroeder, Henry A. *Pollution, Profits, and Progress.* Brattleboro, Vt.: Stephen Greene Press, 1971.

Schumacher, E. P. *Small Is Beautiful.* New York: Harper and Row, 1975.
> Expressive of a common reaction against the defects of the large-scale; and the longing for the virtues of the small-scale.

Schwenk, Theodor. *Sensitive Chaos.* New York: Schocken Books, 1976.

Science 184, no. 4134 (April 19, 1974).
> An issue devoted to energy: people and institutions; policy; economics; oil, coal, gas, and uranium: the developed technology; sun and earth: developing technology.

Science for Development. An Essay on the Origin and Organization of National Science Policies. Prepared by Jacques Spaey. Paris: UNESCO, 1971.

Scientific American 213, no. 3 (September 1965).
> The theme of this issue is cities.

—— 223, no. 3 (September 1970).
> The theme of this issue is the biosphere.

—— 225, no. 3 (September 1971).
> The theme of this issue is energy and power.

—— 231, no. 3 (September 1974).
> The theme of this issue is the human population.

—— 235, no. 3 (September 1976).
> The theme of this issue is food and agriculture.

——. *Science, Conflict, and Society.* San Francisco: W. H. Freeman, 1969.
> Collection of articles from this consistently excellent journal.

Scientific World 19, no. 3/4 (1975).
> Devoted exclusively to "The Role of Scientists and of Their Organizations in the Struggle for Disarmament," a symposium held in Moscow on July 15-19, 1975.

—— 20, no. 2 (1976).
> This issue of the World Federation of Scientific Workers' journal "is mainly devoted to questions of the protection and preservation of the environment." Prepared by the Russian editors, it contains articles on ecology in politics, in Japan, in the biosphere and forecasting, and in developing countries.

Scoby, Donald R., ed. *Environmental Ethics: Studies of Man's Self-Destruction.* Minneapolis: Burgess Publishing Co., 1971.

Sears, Paul B. *Deserts on the March*. Norman: University of Oklahoma Press, 1935.
Man-made destruction of vegetative cover of the earth.
———. *The Ecology of Man*. Eugene: Oregon State System of Higher Education, 1957.
Shapley, Harlow, ed. *Climatic Change: Evidence, Causes and Effects*. Cambridge,
Mass.: Harvard University Press, 1953.
Shepard, Paul. *Man in the Landscape: A Historic View of the Esthetics of Nature*.
New York: Alfred A. Knopf, 1967.
Fertile observations drawn from many areas.
———, and McKinley, Daniel, eds. *Environ/mental: Essays on the Planet as a Home*.
New York: Houghton Mifflin Co., 1971.
"The articles in this new collection are intended to illustrate the scope of
current environmental disorder and the variety of possible perspectives on
it."
———. *The Subversive Science: Essays Toward an Ecology of Man*. Boston: Houghton
Mifflin Co., 1969.
A rich collection of mainly technical essays on a wide range of subjects.
Shibata, Shingo. "Marxismus kontra moderne Eschatologie—Kritik des Klubs von
Rom." *Society and Labor* 21, no. 1-2 (January 1975): 1-58.
Shoji, H., and Miyamoto, K. *Kogai of Japan*. Tokyo: Iwanami, 1975. (In Japanese.)
———. *The Terrible Kogai*. Tokyo: Iwanami, 1964. (In Japanese.)
Sierra Club. *Space for Survival: How to Stop the Bulldozer in Your Own Backyard*.
New York: Simon and Schuster Pocket Books, n.d.
Singer, Peter, and Regan, Tom, eds. *Animal Rights and Human Obligations*. Engle-
wood Cliffs, N.J.: Prentice-Hall, 1976.
Singer, S. F., ed., *Global Effects of Environmental Pollution*. Dordrecht, Holland:
D. Reidel, 1970.
Smith, Frank E., *Conservation in the United States*. 5 vols. New York: Van
Nostrand, Reinhold, 1971.
The Smithsonian Institution. *The Fitness of Man's Environment. Smithsonian
Annual II*. Washington, D.C.: Smithsonian Institution Press, 1968; New
York: Harper and Row, 1970.
An attempt "to relate biologists and anthropologists—students of the
human ecosystem—with planners and architects."
Snow, C. P. *Science and Government*. Cambridge, Mass.: Harvard University Press,
1961.
Dangers of government decisions when the decision-makers are scientifi-
cally ignorant.
———. *The State of Siege*. New York: Charles Scribner's Sons, 1969.
Forecast of possible catastrophic food shortage before 2000. "Peace.
Food. No more people than the earth can take. That is the cause."
Sommer, Robert. *Personal Space: The Behavioral Basis of Design*. Englewood Cliffs,
N.J.: Prentice-Hall, 1969.
Soviet Life, 212, no. 5 (May 1974).
This issue is devoted to "society and nature" and deals with such topics as
the law's protection of nature, clean air, garbage disposal, industry, water, a
clean river, a dam, a city in the desert, rerouting rivers, land, earth studied

from space, forest protection and uses, a strip-mine landscape, air pollution, the desert turned into a garden, insects, and international contacts in ecology.
Soviet Union 290, no. 5 (1974).

An issue devoted to environmental protection in the U.S.S.R.

Spent Nuclear Fuel and Radioactive Waste. Stockholm: LiberFörlag, 1976.

English-language summary of a report of the Swedish Government Committee on Radioactive Waste. See especially the added statement (pp. 91-94) by Dr. John Takman, a committee member from the Swedish Parliament, calling for a broadening of the investigation of nuclear waste to include the military and safety problems connected with the production of nuclear weapons (not only nuclear waste), the dangers of ongoing nuclear weapons tests in the atmosphere, and the stockpiles of thousands of hydrogen bombs and nuclear weapons throughout the world—in short, the problems of armaments and the need for "a broad and effective campaign against the arms race" and for the peaceful uses of atomic energy.

Spring, David, and Spring, Eileen, eds. *Ecology and Religion in History.* New York: Harper and Row, 1974.

Stamp, L. D. *Nature Conservation in Britain.* London: William Collins Sons and Co., 1969.

The State of the Environment 1975. Nairobi: United Nations Environment Programme, 1975.

Brief look at population, food, oceans, energy, raw materials, outer limits, and management of the environment. Based on viewpoints of papers presented to United Nations conferences and on discussions with members of scientific communities.

Stellman, Jeanne M., and Daum, Susan M. *Work Is Dangerous to Your Health: A Handbook of Health Hazards in the Workplace and What You Can Do About Them.* New York: Pantheon Books, 1973.

A theme as old as human history, with new variations in our class society.

Stent, Gunther. *The Coming of the Golden Age: A View of the End of the Progress.* Garden City, N.Y.; Doubleday, Natural History Press, 1969.

Stern, Arthur C., ed. *Air Pollution.* 3 vols.; 2d ed. New York: Academic Press, 1968.

"A very useful standard base line for thought and technology in this genuinely vital field" (Philip Morrison in *Scientific American*).

Sternberg, H. O'Reilly. *Man and Environmental Change in South America.* Berkeley, Calif.: Center for Latin American Studies, 1968.

Stewart, George R. *Not So Rich As You Think.* Boston: Houghton Mifflin Co., 1967.

Stockholm International Peace Research Institute. *The Problem of Chemical and Biological Warfare.* 6 vols. Stockholm: Almqvist and Wiksell International, 1971-1974.

Stone, Christopher D. *Should Trees Have Standing? Toward Legal Rights for Natural Objects.* Los Altos, Calif.: William Kaufmann, 1973.

This eloquent plea, cited by Justice William O. Douglas in his dissent in *Sierra Club v. Morton,* which is included in this volume, originally appeared in *Southern California Review* 450 (1972). The history and meaning of legal rights for persons and things, plus a proposal to confer rights on natural objects and "the natural environment as a whole." Not faced are class questions: What "institutions" make and enforce laws, confer and withhold rights? By what power? If all people participate equally in nature's ecology, should not all have equal right and responsibility toward it? How shall they then displace the power of the institutions?

Storer, John H. *The Web of Life.* New York: New American Library, 1956.
"For everyone interested in the wise use of our soil and water, our forests and wildlife." Shows "how all living things fit together into a single pattern." (Author)

Sutokskaya, I. "Man Is the Chief Concern: Cooperation of Scientists of the USSR and the USA in the Field of Environmental Health Sciences." *Nauka i Zhizn,* no. 1 (1976). In Russian.)

Szafer, W., et al. *Protection of Man's Natural Environment.* Warsaw: Polish Scientific Publishers, 1973.
Collective volume from Poland on environmental control and conservation in relation to human needs and natural laws.

Tao Te Ching. Trans. J.J.L. Duyvendak. London: John Murray, 1954.
Of the dozens of English translations, this is reputed to be one of the best. An early instance of the philosophical formulation of some dialectical and ecological principles.

Taylor, Gordon R. *The Doomsday Book: Can the World Survive?* London: World Publishers, 1970.
———. *Rethink: Radical Proposals to Save a Disintegrating World.* New York: Penguin Books, 1974.

Theobald, Robert, comp. *Futures Conditional.* New York: Bobbs-Merrill, 1972.
Essays by Joseph Wood Krutch, Abraham Maslow, etc.
———, and Mills, Stephanie, eds. *The Failure of Success: Ecological Values Vs. Economic Myths.* Indianapolis, Ind.: Bobbs-Merrill, 1973.

Thoman, Richard S., and Conkling, Edgar C. *Geography of International Trade.* Englewood Cliffs, N.J.: Prentice-Hall, 1967.

Thomas, Lewis. *The Lives of a Cell: Notes of a Biology Watcher.* New York: Viking Press, 1974.
Penetrating and poetic ponderings on what we know and do not about how things mesh and tick—things such as bacteria, songbirds, humpback whales, roses, dogs, organelles, language, Iks, and the Earth itself, "the world's biggest membrane."

Thomas, William A., ed. *Indicators of Environmental Quality.* New York: Plenum Publishing Corp., 1972.

Thomas, William J., Jr., ed. *Man's Role in Changing the Face of the Earth.* 2 vols. Chicago: University of Chicago Press, 1956.
Interdisciplinary studies on man's transformation of his planetary environment, past, present, and future.

Thomlinson, Ralph. *Population Dynamics: Causes and Consequences of World Demographic Change.* New York: Random House, 1965.

Tinbergen, Niko. *Social Behavior in Animals with Special Reference to Vertebrates.*
    New York: John Wiley and Sons, 1953.
Tiselius, Arne, and Nilsson, S., eds. *The Place of Value in a World of Facts: Proceed-
    ings.* Nobel Symposium 14. New York: John Wiley and Sons, 1971.
    See the essay by Harrison Brown, "Resource Needs and Demands."
Toffler, Alvin. *Future Shock.* New York: Random House, 1970; Bantam Books, 1970.
    Dramatic formulation of some major problems.
———, ed. *The Futurists.* New York: Random House, 1972.
    Social critics, scientists, and philosophers and planners reflect on what Toffler
    calls "the emerging 'establishment' within the futurist movement." Included
    are Western scholars like Daniel Bell, Kenneth E. Boulding, Arthur C. Clarke,
    Paul Ehrlich, R. Buckminster Fuller, Bertrand de Jouvenel, Herman Kahn,
    John Wren-Lewis, Marshall McLuhan, and Margaret Mead, as well as the
    Soviet I. Bestuzhev-Lada, the Japanese Yujiro Hayashi, and the Indian M.
    S. Iyengar. Also includes progressives Robert Jungk, Fred L. Polak, and
    Arthur I. Waskow.
Town and Country Planning in Britain. Prepared by Reference Division, Central
    Office of Information, London. Chatham, Kent: W. and J. Mackay, 1975.
Truitt, Willis H., and Solomons, T. W. Graham, eds. *Science, Technology, and
    Freedom.* Boston: Houghton Mifflin Co., 1974.
    A collection of various questions and answers with respect to the economic,
    military, political, moral, and environmental impact of science and technology
    on modern society.
Tsuru, Shigeto. *Political Economy of Kogai.* Tokyo: Iwanami, 1971. (In Japanese.)
———, ed. *A Challenge to Social Scientists: Proceedings of International Symposium
    on Environmental Disruption; March 1970, Tokyo.* Tokyo: Asahi Evening
    News, 1970.
    See the background paper and concluding chapter by the editor, "Environ-
    mental Pollution Control in Japan."
Ucko, Peter J., and Dimbleday, G. W., eds. *The Domestication and Exploitation of
    Plants and Animals.* London: Duckworth; Chicago: Aldine Publishing Co., 1969.
Ui, J. *Introduction to Kogai Studies.* Vols. 1-3. Tokyo: Akishobo, 1971-. (In Japanese.)
———. *Political Economy of Kogai.* Tokyo: Senseido, 1968. (In Japanese.)
Ullrich, Wolfgang. *Endangered Species.* New York: Hart Publishing Co., 1972.
Umweltpolitik. Das Umweltprogramm der Bundesregierung. Stuttgart: W. Kohl-
    hammer, 1974.
    Environmental policies and programs of action in the Federal Republic of
    Germany.
UNESCO. *Use and Conservation of the Biosphere.* Proceedings of the inter-govern-
    mental conference of experts on the scientific basis for rational use and con-
    servation of the resources of the biosphere, Paris, September 4-13, 1968.
    Paris, 1970.
UNESCO Courier, January 1969.
    This issue, devoted to ecology, contains articles by Michael Batisse, René
    Dubos, Jean Dorst, and Frank Fraser Darling.

———, January 1974.

 On "power from the sun and world energy resources."

———, May 1974.

 "An overall picture of the world population situation."

———, July-August 1974.

 Essays on urgent issues.

———, May 1975.

 Essays on "the hunger gap."

Van Melsen, Andrew G. *Science and Responsibility*. Pittsburgh: Duquesne University Press, 1970.

Van Tassel, Alfred J., ed. *Environmental Side Effects of Rising Industrial Output*. Lexington, Ky.: D. C. Heath, 1970.

 Facts, figures, references.

Vernadasky, Vladimir I. *La Biosphère*. Paris: Libraire Félix Alcan, 1929.

 This is the French translation of the Russian *Biosfera* (Leningrad, 1926). Vernadasky developed the idea of the biosphere in Lamarck and Eduard Suess and thus laid the basis for his later elaboration of Le Roy's "noösphere."

———. "The Biosphere and the Noösphere." *American Scientist* 33, no. 1 (January 1945): 1-12.

 Translation of a 1938 abstract and 1943 essay, written during "a stormy time." The conclusion of this creative Soviet thinker is: "We may face the future with confidence. It is in our hands. We will not let it go."

Vickery, Tom Rusk, ed. *Man and His Environment: The Effects of Pollution on Man*. Syracuse, N.Y.: Syracuse University Press, 1972.

Vogt, William, *People! Challenge to Survival*. New York: W. Sloane Associates, 1960.

———. *Road to Survival*. New York: W. Sloane Associates, 1948.

Wagar, W. Warren. *Building the City of Man: Outlines of a World Civilization*. New York: Grossman Publishers, 1971.

Wagner, Philip L. *Human Use of the Earth*. New York: Free Press, 1960.

Wagner, Richard. *Environment and Man*. New York: W. W. Norton, 1971.

Wald, George. "Life and Light." *Scientific American* 201, no. 4 (October 1959): 92 ff.

 Profound dependence of evolving living matter on light.

Wallace, Bruce. *People, Their Needs, Environment, Ecology: Essays in Social Biology*. Vol. 1. Englewood Cliffs, N.J.: Prentice-Hall, 1972.

Wallick, Franklin. *The American Worker: An Endangered Species*. New York: Ballantine Books, 1972.

 From a trade union viewpoint the editor of the United Auto Workers' newsletter has written a popular exposé of the health hazards on the job—the conditions pertaining to the dangerous noise pollution of 17 million workers, 25 million annual injuries, and 4 to 8 million occupational diseases inflicted annually under U.S. industrial capitalism. Wallick updates Marx's documented thesis: that in the capitalist's "own" preserve the workers suffer the most immediate and intense form of a pollution and destruction inflicted on the community at large by its predatory class.

Ward, Barbara. *The Home of Man*. New York: W. W. Norton, 1976.

———. *Space Ship Earth.* New York: Columbia University Press, 1966.

———, and Dubos, René. *Only One Earth.* Harmondsworth, England: Penguin Books, 1972.

———; Runnals, J. D.; and D'Anjou, L., eds. *The Widening Gap: Development in the 1970s.* New York: Columbia University Press, 1971.

Warner, Aaron W., et al. *The Environment of Change.* New York: Columbia University Press, 1969.
  Essays by Isiah Berlin ("The Hazards of Social Revolution"), I. I. Rabi ("The Revolution in Science"), Jacob Bronowski ("The Impact of New Science"), and Loren Eiseley ("Alternatives to Technology").

Warner, Sam Bass, Jr. *The Urban Wilderness: A History of the American City.* New York: Harper, 1972.

Watt, K.E.F. *The Titanic Effect. Planning for the Unthinkable.* New York: E. P. Dutton, 1974.

Weisberg, B. *Beyond Repair: The Ecology of Capitalism.* Boston: Beacon Press, 1971.

———. *Ecocide in Indochina.* New York: Harper and Row, 1970.

West, Felicia E. *Science for Society: A Bibliography.* 4th ed. Washington, D.C.: American Association for the Advancement of Science, 1973.
  Almost 4,000 references, principally of recent date, many annotated, under these heads: reference; science; technology; society; resources and environment; education; health; conflict and population.

Westing, Arthur H., and Pfeiffer, W. W. "The Cratering of Indochina." *Scientific American* 226, no. 5 (May 1972): 20-29.
  A comprehensive report of the extensive damage.

Wheeler, William M. *Essays in Philosophical Biology,* ed. George H. Parker. New York: Russell and Russell, 1967.

Whitehead, A. N. *Science and the Modern World.* New York: Macmillan Co., 1926.
  Speculating about the most general principles implicit in modern physics and biology, Whitehead suggests that the ultimate entities of the universe are not discrete solid atoms but energetic "organismic" activities creatively "prehending" their neighbors. He thus anticipates many principles and concerns of present-day ecology.

Whiteside, Thomas. *Defoliation: What are Our Herbicides Doing to Us?* New York: Ballantine Books, 1970.

Whittaker, R. H. *Communities and Ecosystems.* 2d ed. New York: Macmillan Co., 1975.

Whyte, Lancelot Law. *The Next Development in Man.* New York: New American Library, 1950.
  On the formative tendency in things and thought—a general ecological principle.

Whyte, William H. *The Last Landscape.* Garden City, N.Y.: Doubleday, 1968.
  Use of urban space.

Wieman, Henry N. *The Source of Human Good.* Chicago: University of Chicago Press, 1946.
  The ecology of the "creative event" in the interaction of persons with other persons and with the rest of nature.

Wilson, Carroll L., and Matthews, William H., eds. *Man's Impact on the Global En-*

vironment: Assessment and Recommendations for Action. Cambridge, Mass.:
M.I.T. Press, 1970.

Wilson, Edward O. *Sociobiology: The New Synthesis.* Cambridge, Mass.: Harvard
University Press, 1975.

Organismic, genetic, population, and ecological factors in the adaptiveness
of social groups; the altruistic (and hence adaptive) value of similar genes
(kin selection).

Wilson, William K., and Sholtys, Phyllis, eds. *Environmental Research Organizations
Directory,* 2d ed. New York: Holt, Rinehart and Winston, 1973.

Winn, Ira J. *Basic Issues in Environment.* Columbus, Ohio: Charles E. Merrill, 1972.

Winton, Harry N. *Man and the Environment. A Bibliography of Selected Publications
of the United Nations System, 1946-1974.* New York: Unipub, 1972.

Wolfle, Dael L. *The Home of Science: The Role of the University.* New York:
McGraw-Hill Book Co., 1972.

History of science in the academy, relations of science to government.

Woodburn, John H. *Whole Earth Energy Crisis,* new ed. New York: G. P. Put-
nam's Sons, 1973.

Woodcock, Leonard. "Labor and the Politics of the Environment." *Sierra Club Bulletin*
56 (December 1971): 11-16.

Woods, Barbara, ed. *Ecosolutions: A Casebook for the Environmental Crisis.*
Cambridge, Mass.: Schenkman Publishing Co., 1972.

Twenty-five essays.

Woodwell, George M., and Smith, H. H., eds. *Diversity and Stability in Ecological
Systems.* Brookhaven Symposium in Biology no. 22. Upton, N.Y.: Brookhaven National Laboratory, 1969.

Detailed technical papers from a 1969 symposium. Lacks a philosophical
treatment.

World Armaments and Disarmament. SIPRI Yearbook 1974. Cambridge, Mass.:
M.I.T. Press, 1974.

Similar yearbooks for 1968-1969, 1969-1970, 1972, and 1973, prepared by
the Stockholm International Peace Research Institute, have been published.
This volume contains the latest data on the subject.

World Conference on Development. Helsinki: World Peace Council, 1976.

Report of the October 1976 conference held in Budapest, "the largest international gathering of representatives of public opinion ever held on the subject of development," aimed at mobilizing public support for decisions and
resolutions of the United Nations and its agencies in the battle against neo-
colonialism and for economic development.

World Conference to End the Arms Race, for Disarmament and Detente. Helsinki:
Continuing Liaison Council of the World Congress of Peace Forces, 1976.

Record of September 1976 Helsinki conference of more than 500 delegates
representing 90 countries and widely diversified groups and political views.

Yermolenko, Dimitry. *International Relations in the Era of the Scientific and Technological Revolution: Problems and Prospects.* Moscow: Novosti, 1973.

See pp. 72-84 for a discussion of environmental protection.

Index

About the Editor

Howard L. Parsons, professor of philosophy and chairman of the department of philosophy at the University of Bridgeport, specializes in social philosophy. He has written articles for such journals as *The Journal of Philosophy*, *The Journal of Religion*, and *Philosophy East and West*. His previous books include *Humanism and Marx's Thought; Marxism, Revolution, and Peace; Man East and West;* and *Dialogues on the Philosophy of Marxism* (Greenwood Press, 1974).